Critical Perspectives on Open Development

The MIT Press—International Development Research Centre Series

Open Development: Networked Innovations in International Development, edited by Matthew L. Smith and Katherine M. A. Reilly

Public Access ICT across Cultures: Diversifying Participation in the Network Society, edited by Francisco J. Proenza

Shadow Libraries: Access to Knowledge in Global Higher Education, edited by Joe Karaganis

Digital Economies at Global Margins, edited by Mark Graham

Making Open Development Inclusive: Lessons from IDRC Research, edited by Matthew L. Smith and Ruhiya Kristine Seward

Critical Perspectives on Open Development

Empirical Interrogation of Theory Construction

Edited by Arul Chib, Caitlin M. Bentley, and Matthew L. Smith

The MIT Press
Cambridge, Massachusetts
London, England

International Development Research Centre
Ottawa • Amman • Dakar • Montevideo • Nairobi • New Delhi

Published by the MIT Press.

A copublication with
International Development Research Centre
PO Box 8500
Ottawa, ON K1G 3H9
Canada
www.idrc.ca / info@idrc.ca

The research presented in this publication was carried out with the financial assistance of Canada's International Development Research Centre. The views expressed herein do not necessarily represent those of IDRC or its Board of Governors.

This book was set in Stone Serif and Stone Sans by Westchester Publishing Services.

Library of Congress Cataloging-in-Publication Data

Names: Chib, Arul, editor. | Bentley, Caitlin M., editor. | Smith, Matthew L., editor.
Title: Critical perspectives on open development : empirical interrogation of theory construction / edited by Arul Chib, Caitlin M. Bentley, and Matthew L. Smith.
Description: Cambridge, Massachusetts : The MIT Press, [2020] | Series: The MIT Press--International Development Research Centre series | Includes bibliographical references and index.
Identifiers: LCCN 2020024468 | ISBN 9780262542326 (paperback)
Subjects: LCSH: Information technology--Economic aspects. | Information technology--Social aspects. | Information society. | Information commons. | Open source software. | Economic development.
Classification: LCC HC79.I55 C75 2020 | DDC 303.4833--dc23
LC record available at https://lccn.loc.gov/2020024468

ISBN: 978-1-55250-596-0 (IDRC e-book)

Contents

Acknowledgments

This research was conducted as part of the Strengthening Information Society Research Capacity Building Alliance (SIRCA) III program and is funded by Information and Networks in Asia and sub-Saharan Africa, a partnership between the United Kingdom's Foreign, Commonwealth and Development Office (FCDO) and Canada's International Development Research Centre (IDRC), under the Networked Economies program at IDRC.

We would also like to acknowledge two SIRCA members, Yvonne Lim and Sandy Pek, whose contributions were critical to this volume's success.

At IDRC, we thank Catherine Bienvenu for copyediting the manuscript, and Nola Haddadian, publisher, for providing invaluable assistance in publishing our manuscript. We also appreciate the contributions of MIT Press's Gita Devi Manaktala, editorial director, and María Isela García, acquisitions assistant, to the publication of this volume.

groundbreaking work that will define not only our program but the field of open development in and of itself.

As we look to SIRCA's future, we have learned a great deal from the SIRCA III experience. Global challenges related to changes in the geographies of poverty and population migration, as well as severe social and gender inequalities, are now markers of our time. Openness may indeed mitigate some of these global challenges, but it seems that a more targeted, aggressive research agenda, focusing on reducing social inequalities, is needed. We encourage future research programs to focus on reducing disparities surrounding gender and technology specifically. There is much to look forward to.

About the SIRCA III Program

The SIRCA III program has evolved significantly since its first and second iterations. The third program was funded by a research grant award focused on developing crosscutting theoretical frameworks in the area of open development. SIRCA III funded projects were led by twelve distinct teams of two to three senior researchers and/or practitioners, resulting in empirical investigations in developing countries globally.

SIRCA III had two phases in its research design—a theoretical phase (Phase I) and an empirical phase (Phase II). During the first year, senior research teams were awarded funding to develop a cross cutting theoretical framework. This framework was tested during the empirical phase of the program in the second year. Additional researchers were selected, through a second call for proposals, to conduct the empirical research in close collaboration with the theoretical research teams.

This edited volume explores cross cutting open development themes and raises issues about the legitimacy and overall purpose of open development. It represents a remarkable evolution in the conceptualization and application of digitally enabled openness to influence positive social transformation. The volume pushes past a theoretical level of engagement with open development and puts the SIRCA III authors' ideas and theories to the test. In this groundbreaking research, cross cutting themes were empirically tested in Asia and sub-Saharan Africa, and the authors reflect on how to improve proposed theoretical lenses. This volume therefore combines theoretical views with their practical application. The authors likewise critically

Preface

The Strengthening Information Society Research Capacity Alliance
program ran for over a decade. What began as an intensive men
program for Asian scholars soon expanded to include members from
the globe. Driven by a passion for social change through the applica
information and communications technology (ICT) in developing
tries, SIRCA scholars have demonstrated outstanding influence. They
shifted ICT4D (ICT for Development) discourses and taken up lead
roles within their institutions. Now, following the end of SIRCA's
iteration, we have successfully developed and tested a more concent
research model focused on a central objective—to build the field of
development. Focusing on the impact of digitally enabled openness
reducing global inequalities enables SIRCA to bring our transformed
gram design into effective action.

In 2015, we began theorizing open development by centering on wheth
how, for whom, and in what circumstances does the free, networked, pub
sharing of digital (information and communication) resources contribu
toward (or not) a process of positive social transformation. Six teams
leading scholars developed white papers, which were released to the publ
to enable external researchers to build research proposals for investigation
An additional six teams of empirical researchers were selected in 2016 to
put our theories to the test.

This simultaneous theoretical and empirical approach, with minimal conflicts of interest between the two sets of scholars, is novel in ICT4D research. SIRCA's theory-driven and empirically tested research model has the potential to serve as a gold-standard practice in ICT4D research. The SIRCA III program has culminated in a book that demonstrates the reflect on such an approach, yielding a valuable source of reference for this emerging area of research. The book also proposes a new model of research within the area of ICT4D as a means of reducing the growing disparities between the potential and realities of how and whether digitally enabled sharing contributes toward a process of positive social transformation.

This volume deepens our understanding of open development in three significant ways:

1. It focuses on generating cross cutting theory that is widely applicable and contextually relevant. This contribution departs from the dominant functionalist approach in the field. It also centers on themes that have broad utility across a range of practice domains and institutions.
2. It emphasizes a transformational lens, such that power, marginalization, and the socially embedded nature of open development are core elements within theoretical development. This addresses a need in open development research to not only identify structural inequalities within development processes but also constructively address them at a fundamental level.
3. It takes a two-stage approach to confirm, test, deconstruct, modify, and improve proposed theory. Our approach offers significant empirical insights into open development by examining new and mature initiatives in four countries. It also enables a critically reflexive approach to theory building, which is grounded in realities faced by poor and marginalized people.

Arul Chib, Caitlin M. Bentley, and Matthew L. Smith

1 Openness in International Development

Caitlin M. Bentley, Arul Chib, and Matthew L. Smith

Introduction

This book is about deepening our understanding of how, to what extent, and in what contexts openness contributes to a process of positive social transformation. By openness, we mean sharing free, public, networked information and communication resources. Specifically, the book focuses on crosscutting themes that mediate *open processes* in international development contexts.

Over the past two decades, open processes around the sharing, use of, and collaboration with digital information and communication resources have emerged in earnest. Open processes typically offer license to use, reuse, and modify resources for free and do not impose access restrictions. As Tim O'Reilly (cited in Macmanus 2004, n.p.), creator of the participatory web, stated, "The network is opening up some amazing possibilities for us to reinvent content, reinvent collaboration." Few innovations of the past quarter century epitomize the transformative potential of networked technologies such as open processes. Producing open source software (van Reijswoud and de Jager 2008), sharing open and linked data (Powell, Davies, and Taylor 2012; United Nations Global Pulse 2012), producing crowdsourced knowledge (Saif et al. 2009), sharing open access publications (Nyamrjiah 2009), creating open educational resources (Percy and Van Belle 2012), and using Web 2.0 tools (Sadowsky 2012) are examples of open processes with widespread potential to facilitate positive social transformation.

The popularity of *openness* among the international development community reflects a desire for a more equal, just, efficient, and ecological world. However, open processes are sometimes implemented poorly and can have contradictory effects. Take, for example, Transparency International's open initiative to reduce police and government corruption in Cambodia. "Have you ever been forced to pay a bribe?," reads the website of the Cambodian chapter of Transparency International, "Bribespot allows you to pinpoint your encounters with public officials easily and anonymously, using either the mobile app or the website" (Transparency International Cambodia 2018). The Bribespot app positions itself to enable citizens to have a voice, to address corruption, and to create a more responsive government through reporting bribes. By aggregating reports and displaying this information on a map, policymakers can obtain a better understanding of problematic bribery hotspots. This sort of innovation is a symbol of change, yet it relies on deeply embedded technological, deterministic assumptions because an app does not automatically change citizens' propensity to share bribe reports and does not guarantee their security if they do. As a result, Paviour (2016) reported that there have been only sixty bribes reported since 2014 and that the app is not being used widely. While open processes offer opportunities for positive social transformation in ways that were not possible in the past, there is still a need to understand how and why these transformations occur and to interrogate our assumptions and the contextual conditions when these do not.

The main purpose of this book is to establish and test theory that actively engages with ideas and practices of open development and that can be contextualized and practiced widely. Through investigations of crosscutting themes—trust, learning, critical capabilities, and stewardship—this book explores how theoretical frameworks can be applied to deepen our understanding of open development. Empirical reflections in this book then give accounts of how open development discourses emanating from traditional development actors, such as governments, international aid institutions, and researchers, shape ideas and practices of open development. The authors contribute key insights into how and why power, practice, and the institutionalization of open processes can reduce social inequalities and empower poor and marginalized people.

This book is intended for researchers studying open development, for practitioners implementing it, and for policymakers who decide which open initiatives to fund and how it should be done. This book offers empirically tested theoretical frameworks for understanding key processes of open development through situated, critical, and sociotechnical lenses. It contributes to our understanding of openness and information and communications technologies (ICTs) in development contexts by bringing together insights from multidisciplinary perspectives and interrogating concepts in context. For policymakers, this book provides a statement in favor of adopting a more contextually informed and responsive approach to managing open initiatives. It makes concrete suggestions to help open initiatives live up to their promise and focuses on enabling the millions of people living in poverty across the globe to benefit from these initiatives.

Delimiting the Boundaries of Open Development

The study of *openness* and *open development* in the context of international development is a recent phenomenon. The global observance of a *network society* has spurred a diverse literature base, which identifies, analyzes, and delimits openness heterogeneously. Moreover, the purpose of open development research differs significantly between technical and social disciplines, making it problematic to suggest a common research agenda. Cutting across these differences and ambiguities is the idea of networked information and communication resources, which are characterized by the freedom to access, reuse, and redistribute qualities. Such qualities enable sharing, reusing, distributing, and repurposing these resources through distinct networked activities. Often, these networked activities appear in response to failures in government and market inefficiencies in developing countries. These activities exist independently of and in conjunction with traditional development activities and are enacted by both traditional development actors and those who do not conform to or abide by the rules of public, private, or nongovernmental institutions.

Much debate has centered on problematizing the conceptual domains of *openness* and *open development*. For openness, many definitions have been proposed across social and technical sciences and the humanities. Pomerantz and Peek (2016) referred to "Fifty shades of open" to highlight the multiplicity of meanings and criteria of openness, many of which existed prior to the Internet and digital technologies. Recently, theories of openness have focused on the qualities and characteristics of openness, often distinguishing between what is open and what is closed, or on licenses, legal or technical frameworks relating to digital information, and communication resources (Berners-Lee 2006; Open Knowledge International 2016). Broadly, openness means (1) free access to resources, (2) unbounded use potential, (3) lack of access or use restrictions of open resources or processes, and (4) favored use of open resources when creating new ones (Pomerantz and Peek 2016).

However, in development, most networked activities we classify as open do not fit the mold of what may be considered open in the global arena. For *open development,* it is necessary to consider not only precisely what openness is but also what it contributes. For instance, data that is shared publicly but that may not be in a standard technical format or have an open license may still have a significant effect on informing the public. Whether and how this information sharing has an impact on development depends much more on how and why stakeholders enact openness than on the conditions of openness. Our working definition of open development therefore clarifies what is meant by openness and includes a normative dimension regarding what we intend openness to contribute. Open development is the free, public, networked sharing of information and communication resources toward a process of positive social transformation.

A key advancement in the concept of openness was made by Smith and Seward (2017). They argue that we stand to learn much more about openness in development when we observe openness in practice. Specifically, the free, public, networked sharing of information and communication resources happens through one (or more) of three open processes:

- *Open production* This concept expands the boundaries of who can participate in a production process through the practices of peer production or crowdsourcing. The key features that make these two models of production open are that (1) participation is both free and voluntary and (2) participation is nondiscriminatory with respect to who can participate (considering the locational and interest boundaries of whatever is being crowdsourced).
- *Open distribution* This is the sharing of digital content for use by others, typically on the Internet. The key features are (1) content that is shared for free and (2) content that is nondiscriminatory with respect to who can access and use it.
- *Open consumption* This is the set of uses of freely shared digital resources. For free and open source software (FOSS), for instance, these uses are operating, copying, distributing, studying, changing, and improving the software. For open educational resources (OERs), these are the 5Rs: retaining, reusing, revising, remixing, and redistributing (Wiley 2014). The Hodgkinson-Williams (2015) framework provides a useful extension of

Wiley's 5Rs with variations of the Rs and an additional sixth practice, that of creation. Note that redistribution refers to both consumption and sharing (Smith and Seward 2017).

These open processes serve as a blueprint for researchers and practitioners to recognize openness in context. Table 1.1 further elaborates on the key characteristics and examples of each open process. Smith and Seward (2020) further elaborate on the *openness as praxis* perspective. They argue that focusing on how and why openness is practiced enables greater flexibility in acknowledging the importance of the contexts shaping the meanings and outcomes of these processes. The practice perspective also emphasizes examining how and why socioeconomic, political, and cultural factors influence open processes. The authors of the chapters in this book engage with this perspective on openness, stating when they adopt an alternative view.

Table 1.1
The three main types of open processes, connected open practices, key characteristics, and examples

Open process	Open practice	Key characteristics	Examples
Open production	Peer production	*Decentralized* governance Nondiscriminatory Voluntary contributions Free to participate	Open source software production, Wikipedia, open legislation
	Crowdsourcing	*Centralized* governance Nondiscriminatory Voluntary contributions Free to participate	Open innovation, citizen science, Ushahidi, ICT-enabled citizen voice
Open distribution	Sharing, republishing	Nondiscriminatory Nonproprietary Typically via platform	Open government data portal, OER portal (e.g., Khan Academy), open access journals
Open consumption	Retain, reuse, revise, remix	Freedom to use	Translating educational materials, taking a massive open online course (MOOC), intermediary visualizing open government data

Source: Smith and Seward (2017).

Conversely, clarifying what we mean by development or positive social transformation is a more complicated task. This is especially true because the authors of this book do not take a uniform approach to social transformation. The next section provides a general overview of the different approaches to development that can be observed in this area and how openness can potentially contribute. This overview is then contextualized by the theoretical approaches to openness and social transformation found in this book.

Obscuring the Boundaries: What Kind of Social Transformation Are We Talking About?

Open development is a relatively new field of research and practice. It has emerged and developed (and may become obsolete) as a response to the global development climate within which it operates. Hence, the first task in this book's discussion is to understand the relationship between open development and the wider international development arena. This book illustrates the interplay between how dominant and alternative perspectives of *ICT*, *openness*, and *development* may impact the theory and practice of *open development*. This section outlines a brief overview of the dominant theoretical influences underpinning open development since its emergence around 2008. It draws on development theory to outline perspectives on social transformation that the authors engage with. This is intended to help the reader engage with the different positions the authors demonstrate throughout the book.

While there has been increasing interest in open development, limited attention has been paid to its relationship to development theory. Instead, many scholars have focused on the ideological dimensions of openness, concentrating on how the processes and characteristics of open innovations broaden the potential for societal change (Bentley and Chib 2016). However, this ideological dimension of openness does not take into account how openness is also embedded and embodied within existing development discourses. Furthermore, broader discourses outlining how ICTs have been integrated into development thinking are sparse and disjointed. To provide a general overview for the reader, we outline the role of technology in development, noting differences between technology and open processes throughout.

Development theory and practice began decades ago. However, since open development emerged around 2008, we focus our overview on contemporary theory and practice since that decade. We have adapted Willis's (2011) summary of the main development approaches and ideas to incorporate the role of ICT within each frame because it succinctly explains the dominant ideologies chronologically. We have added aid effectiveness and inequality to the discussion to reflect more recent development trends since 2010. This overview sheds light on the variety of interpretations and underpinning theory that can be observed within open development.

Each development approach has its own history, academic debates, and lessons learned. As Willis (2011) remarks, there are many ideas and practices that operate simultaneously, but those that are popular are typically heralded by the most powerful actors. Table 1.2 indicates that ICTs have been critical components in a number of approaches to development, and they continue to be sources and topics of debate. Sachs et al. (2015), for example, positioned ICTs as a main enabler of development and key to achieving the Sustainable Development Goals of the United Nations. Unwin (2017), in contrast, argued that ICTs have overwhelmingly been linked to ideas and practices of development related to economic growth, which have now brought about more inequality than the world has ever experienced. Ultimately, if ICTs are to truly support development, researchers must focus much less on the technologies themselves and more on the processes and approaches that can protect and empower poor and marginalized populations. Such contrasting views are also found within the area of open development, as the next section demonstrates.

A Brief History of Openness and Development Theory

In the past decade, a diverse range of sources have been increasing emphasis on openness within the development discourse, in many cases arguing that openness creates benefits within development aid relationships and in public life (Buskens 2013; Cyranek 2014; Reilly and Smith 2013). It is important to distinguish where interests in establishing openness in development come from, as this could indicate what development outcomes are actually sought. Many of the development strands outlined in table 1.2 are observable within open development at present.

Table 1.2
ICT in development

Main development approaches	Main development idea	Role of ICTs
Neoliberalism	Increased engagement with concepts of globalization	ICTs are cost-effective, scalable tools to support efficient democratization and privatization in service delivery.
Postdevelopment	Ideals about *development* represent a form of colonialism and Eurocentrism. Should be challenged from the grassroots	ICTs are typically agents of modernization, which destroy local cultures and economies. They can also be used to preserve Indigenous cultures and histories.
Sustainable development	Need to balance needs of current generation against environmental and other concerns of future populations	ICTs are key sources of technical innovation equipped to solve environmental problems and mitigate disaster. Humans alone are not capable of reversing the effects of climate change.
Grassroots approaches	Importance of considering local context and Indigenous knowledge	Participatory designed ICTs are viewed as sources of empowerment, and community-based approaches to knowledge sharing are often seen as positive for social change.
Rights-based development/development as freedom	People should have the freedom to lead the lives they have reason to value. Basic human rights should be universally guaranteed.	Internet access is adopted as a universal human right, and initiatives to make the Internet more accessible and affordable continue. Capacity gaps in Internet infrastructure play important roles in mediating capabilities.
Inequality	Greater awareness of the ways in which globalization and neoliberalism advantage the wealthiest 1 percent of the global population and erode the middle class in many of the so-called developed countries	ICTs are acknowledged as pervasively responsible for reinforcing global inequalities and further entrenching dependencies.
Gender and development	Greater awareness of the ways in which gender is implicated in development	ICTs are acknowledged as sources of gender inequality, violence, and harassment. Conversely, they can be used to support initiatives to empower women.
Aid effectiveness	Greater focus on recipient nation ownership of aid, transparency, governance, and demonstration of certain kinds of results as evidence of achieving development	ICTs, specifically new open data techniques, are considered a key way to support transparency and macro-level learning about what works well for development.

Source: Adapted from Willis (2011, 28).

Before the twentieth century, development was considered something that just happened. Because of the influence of evolutionary theories generated within the natural sciences, it was believed that societies developed on their own and consequently not as a result of human influence (Latouche 1988). Cowen and Shenton's (2003) distinction between "big D" and "little d" development is still helpful to discern when development is thought to just happen and when it happens through intervention. Indeed, much of the early network society literature positioned openness as an immanent development, or "little d" development, that is the result of societal evolution and technological innovation. The latest of these developments seems to be blockchain technologies, which may completely disrupt centralized banking and macroeconomic policymaking (Zambrano 2017). Therefore, to date, a majority of research on openness in the context of development has sought to learn how to take advantage of open processes by focusing on the creation of new tools and techniques to solve development problems (Bentley and Chib 2016). Smith and Seward (2020) offer a thorough discussion of how openness has evolved across different domains of practice outside the development sector, which may heavily influence how open development is conceptualized and practiced. However, addressing severe global differences and inequalities or the perspectives of poor and marginalized people has not been a main focus of these studies or domains.

In contrast, the development sector is known for practicing "big D" development, and it has a long history and many competing visions (table 1.1). The neoliberal approach to development has focused on openness because of its potential to encourage innovation and good governance. Initially, it seemed that openness was meant to reduce economic inefficiencies caused by the commodification of knowledge and information. Heller (2008) has argued that scarce resources can be underutilized if too many private owners control complementary resources. For instance, if intellectual property rights are too heavily controlled, economic and social inefficiencies can occur. Heller's (2008) research illustrated that when drug patents are owned by numerous pharmaceutical companies in the United States, incentives to innovate are diminished. This leaves potentially disease-curing drugs abandoned when the cost of patent royalties outweighs potential profits.

The World Bank Institute (2014, 1) has also increasingly used innovation as a backing for its open agenda, stating that "there is growing recognition that governments acting alone cannot provide public services to all of their citizens. They need partners from civil society, commercial enterprises, and private non-commercial actors including social entrepreneurs to complement, support, and create new business models for the delivery of public goods and services." However, within the World Bank's argument, innovation is less about governments needing to fix economic inefficiencies that impede innovation and more about finding ways to increase government transparency and to support market-led responses for public service provision. Bates (2014) has argued that information policies have played a key role in pushing forward neoliberal governance objectives, such as the marketization of public services and privatization of public assets in the United Kingdom. She also argued that the openness agenda has served to bolster politicians' reputations (Bates 2014). Her findings point to a need to be cautious and critical of openness and to be clear about the outcomes that are sought.

The neoliberal approach emphasizes economic growth and making existing economic and social structures (such as state governments and private corporations) more efficient in global development. Deseriis's notion of technolibertarianism resonates with these neoliberal development objectives because it "is often referred to as an ideology that combines a blind faith in technological progress and free-market economics with a deep distrust in statist, bureaucratic, and hierarchical forms of authority" (Deseriis 2017, 442). In practice, this philosophy has translated into a host of international programs focused on good governance objectives, such as the Open Government Partnership, which started in 2011 (see in this volume Mungai and Van Belle, chapter 5; Moshi and Shao, chapter 12). Such programs and tenets have contributed two significant attitudes about openness in development: the belief that openness through networked technology is a change agent of major importance for initiating development as economic growth and that openness provides a means to foster good governance. Thus, the neoliberal approach bestows capitalist and democratic governance systems with the highest value and views openness as supporting these interests.

Furthermore, openness has been an important aspect of aid effectiveness discourses. Some donors, along with civil society organizations and private foundations, have campaigned to improve transparency and accountability in development aid spending through openness: "Sharing aid information more effectively will ultimately enable stakeholders to build up a richer picture—by allowing more information to be aggregated and by allowing innovation in the way this information is represented and queried" (Gray et al. 2009, 3). Publish What You Fund (2015) campaigned on the platform that donors and governments in many cases have information but find it difficult to find information that is needed in order to make informed decisions concerning how and where to spend development aid money. Recipient governments are unable to budget their own resources properly, as they cannot easily get a concrete estimate of how much money is coming into the country and for which purposes (Moon and Williamson 2010). The International Aid Transparency Initiative (IATI) was created to *unlock the potential* of new technologies to make this information more useful (Gray et al. 2009). This purpose primarily seeks to make the current aid system more efficient and productive.

These discourses on productivity and efficiency are situated within a framework of development that remains tied to notions of *underdevelopment*, which refers to basing development on making underdeveloped countries more like their richer, so-called developed, counterparts. Additionally, this perspective includes the idea that there are development problems that can be solved on a universal scale by drawing on theory and practices of scientific inquiry, with technical assistance as a main feature. An enduring critique of this perspective is that cultures of so-called underdeveloped countries, in particular Indigenous cultures, are regarded as barriers to development (Crush 1995). They are barriers since they hinder formation of the efficient, consumptive cultures heralded by this perspective as the ultimate stage of human development.

It is not surprising then that a number of academics, such as Rahnema and Bawtree (1997) and Escobar (1995), have argued that development encourages individuals to internalize others' perceived inadequacies of themselves and their societies. For instance, poor and marginalized populations are encouraged to obtain standard education, which historically has devalued Indigenous knowledge. Students develop into citizens untrained to resist attacks on their cultures and practices. In school, they also learn all the capitalist doctrines within which progress is equated with economic growth, and they learn that finding a job is what is needed to earn a living. Tsing (2016) has therefore argued that when forest-dwelling populations in Indonesia were forced to relocate through resettlement programs, they had to learn how to be poor. These populations had to give up their livelihoods and so-called *backward* ways of living in order to assume inferior positions within society, all because logging and mining companies needed to make a profit in the name of development.

Postdevelopment, postcolonial, and feminist theorists have focused on the power structures of society, leading us to question the meanings and practices of development. These theorists view development as a discourse, with its own set of actors, regularities, and outcomes that take place within a complex global system. To these writers, analyzing how discourses are constructed and how they operate while situated within a specific context is what is needed to understand development. According to this perspective, development cannot be adequately represented by economic principles, such as productivity and efficiency, alone. Analyzing development discourse implies revealing how researchers, practitioners, and policymakers speak about development, and how their practices and languages change over time. It also involves bringing geopolitics into the discussion in order to address how culture, time, and place influence practices of development.

Within this context, at least three approaches to open development stand in opposition to models overly focused on economic growth, productivity, or efficiency. The first approach aligns with Amartya Sen's notion of development as freedom, which argues that people should have the right to lead the lives they have reason to value. Reilly and Smith defined open development as the potential to "expand individual *freedoms* through more participatory processes and by enhancing voice, as well as expand people's *capabilities* through increased access to resources, in particular digital information and connections to people and all they bring" (Reilly and Smith 2013, 32). Within this strand of open development, openness is valuable because it enables communities to make information not only accessible but also reusable and modifiable. Communities that develop common technological tools and knowledge resources may be better positioned to support a wide variety of needs and wants, thus building an infrastructure for freedom. Networked technologies could also enable communities to share across geographical locations so that they may work together collaboratively to fulfill a greater diversity of needs. Smith and Seward (2020) investigate a range of programs that sought to implement this open development approach. Their book chronicles major challenges to include poor and marginalized populations. Thus, there is still a need to understand how open development may be practiced more effectively.

The second approach to open development seeks to address structural inequality by including more political concepts of development and draws on participatory governance ideals. For instance, Singh and Gurumurthy's (2013) idea of a *network public* is an institutional ecology that is not only accessible to all people but is owned and operated jointly. Similarly, Restakis et al. (2015) proposed *generative democracy* as an alternative that emphasizes citizen engagement and the incorporation of social knowledge in public service systems. These notions view open development as a means to claim citizenship rights, develop political processes, and increase political agency of poor and marginalized actors. For this political strand of open development, openness builds on notions of deliberative democracy (see Dryzek 2000). This idea is centered on the notion that people should not only have access to open resources but should also be able to deliberate issues of communal interest, debate validity claims, and have a say in deciding which ethical actions to take (Habermas 1987). The purpose of openness is to build such communal decision-making processes into the practices and institutions of development, but it is still not clear how openness may contribute toward these aims.

The third approach to open development is more radical. For Bauwens and Kostakis (2014), open processes create common pools of knowledge that can potentially serve the interests of many groups and individuals inclusively. Yet openness becomes a parody when start-ups and large multinationals exploit and capitalize on these commons beyond what they contribute back. Similarly, alternative economic models to capitalism, such as those practiced within the solidarity economy (e.g., cooperatives, FOSS, fair trade, or recycling), do not by themselves have the potential to remedy unsustainable and exploitative capitalist practices. Alternative models and institutions eventually need to compete with private actors and corporations, which incentivizes them to operate within the capitalist paradigm. For instance, to remain competitive, worker cooperatives may need to develop and maintain their own intellectual property beyond their contributions to common pools of knowledge and resources. Therefore, Bauwens and Kostakis (2014) argue for *open cooperativism* as a radical way to completely restructure politics and society. They argue that convergence of the previous two approaches to openness—greater freedom through open production and common ownership and governance—is what is needed to transition toward a commons-oriented economy and society. However, it remains unclear how and whether such a transition will realistically take place.

Putting Openness and Development Theory into Practice: Toward Critical Perspectives on Open Development

As outlined, openness has a wider range of roots in development theory and practice than is usually recognized. In practice, political agendas and institutional policy may have had significant influence over how openness tends to be enacted in support of productivity and efficiency. This is a key point since much of the rhetoric of openness has emphasized freedom, participation, and inclusion in development. Yet, this rhetorical focus has tended to neglect the diversity of open development approaches occurring and has failed to address theory and practice perspectives that may need to be actively confronted. In contrast, we argue that for open approaches to be truly transformative requires critical engagement across the spectrum of development approaches.

Work to *transform* existing "big D" development practices and institutions is central to many approaches to open development. More ambitiously stated, during the transformation process, there is an emphasis on rectifying social and political relations, institutional structures, and societal inequalities that cause poverty and marginalization. However, there is still a need to define precisely what *transformation* means in open development. For instance, does it mean focusing on either local or international contexts, combined with specific populations or certain development outcomes? Realistically, it is overly ambitious to expect open development initiatives to transform existing power relations rapidly. For example, sharing information publicly through the web is unlikely to transform the infrastructures and resources that keep poor and marginalized people from participating in open development initiatives in the first place. Such a limited, constrained, and short-term view of open development may lead us to abandon the field altogether.

In contrast, we argue that there is much to be gained from deepening our understanding of open development from a critical perspective for three reasons. First, openness, like development, is not neutral and is keenly contested. When researchers uncover the ideological tenets that link openness explicitly to a theory of development, there may be a greater chance that interests underpinning openness may come to light. Second, practicing openness has the potential and capacity to distribute knowledge and power across contexts, albeit in good and bad ways. There still remains a need to uncover how and why changes in such distributions take place. Critical perspectives taking into account both the positive and negative effects of openness may be better positioned to understand open development. Moreover, the freedom and political approaches to open development demonstrate a need to explore transformation at individual and local levels in comparison to the institutional and structural elements that influence open practices. We argue that examining the practices and institutionalization of openness—drawn in part from critical perspectives of development—provides a useful way to build theoretical frameworks that work strategically toward positive transformation. The next section further outlines the approaches adopted by the authors of this book.

Two Overarching Perspectives on How Openness Contributes to Positive Social Transformation: Overview of the Book

This volume is the result of a three-year Strengthening Information Society Research Capacity Alliance (SIRCA) research program focused on generating crosscutting open development theory. The program consisted of two distinct phases. In the first phase, in 2015, SIRCA launched an open call for research proposals to develop crosscutting theoretical frameworks. The only condition placed by the program was that research teams had to consist of two principal investigators with different disciplinary backgrounds. Six multidisciplinary research teams were selected based on the crosscutting potential of the theme and the quality of the proposal. This volume outlines these theoretical frameworks, which engage with the themes of stewardship; trust; situated learning and identity; understanding inequalities; critical capabilities; and increasing transparency, participation, and collaboration within institutions.

As a result of progress within the field of open development, the teams shared a common goal of building theoretical lenses that were both practical and widely applicable across domains. For instance, Smith and Reilly (2013) built on earlier work to define open development, and the theoretical contributions in their book outlined robust potential for resolving challenging development problems. Since then, a wealth of research has been published, with limited evidence of impact (Bentley and Chib 2016;

Bentley, Chib, and Poveda 2018). In the first volume of this series, *Open Development: Networked Innovations in International Development* (Smith and Reilly 2013), the authors engaged with empirical evidence in order to better understand how openness may be governed toward more inclusive developmental impacts within various domains of practice (such as open government data initiatives, open science, and gender mainstreaming). In contrast, this volume considers important issues that span across domains with the intention of improving the way openness is practiced and institutionalized toward a process of positive social transformation.

To avoid *theory for theory's sake* as well as theory merely for *critical deconstruction*, we devised a second stage to empirically interrogate the theoretical frameworks. In May 2016, the six theory teams published early versions of their theoretical frameworks for public review. A second call for proposals was launched to invite submissions for empirical investigation of the theoretical frameworks. Six empirical projects across the United Republic of Tanzania, the Democratic Socialist Republic of Sri Lanka, the Republic of Kenya, and the Republic of India were selected. We invited proposals to test frameworks across domains of practice. Furthermore, the second round imposed certain conditions on project submissions for the research to take place in certain priority countries identified by the program funders. Moreover, the research timeline also constrained what was feasible to achieve within the time available. Our hopes that each theory would have empirical testing across domains were not feasible. Therefore, each empirical study was conducted with a focus on one context and typically one domain of practice. Nevertheless, the dialectical process between theory and empirical teams greatly improved the practical relevance of the theoretical frameworks.

This book is structured in two parts, pragmatic approaches to open development and coevolutionary perspectives on open development, each with a series of empirical reflections that respond to the proposed theoretical frameworks. These parts reflect the different theoretical approaches that emerged through the SIRCA process. All the contributors analyze the complex ways in which development—in all its various practices and institutions—is both hindered and helped by *openness*. The authors likewise interrogate the meanings and practices of openness that permit empowering forms of development or processes of positive social transformation. Thus, the chapters are not simply a collection of theoretical frameworks but instead engage collectively in addressing power and institutional inequalities within open development. The next two sections highlight how the authors achieve this.

Part I: Pragmatic Approaches to Open Development

This volume opens with three chapters that suggest three very different approaches to understanding crosscutting themes of open development. Taken together, these chapters offer possibilities to pragmatically investigate stewardship, trust, and situated learning and take as embedded within them power differentials between the actors involved. These theoretical frameworks offer constructive critique and practical suggestions to move debates and progress forward. Thus, pragmatic approaches to open development are:

- Context driven
- Practice based
- Strategic
- Bridge practices and strategic functions and purposes of openness

These frameworks can potentially be applied to study a wide range of development approaches and can be adapted to many different contexts and initiatives. Bridging practices and strategic functions means that it is up to the actors involved to use the frameworks to build deeper understandings of data stewardship, as well as trust and situated learning outcomes, and to carry these forward.

Reilly and Alperin (chapter 2) outline how open data for development could be further enriched by analyzing stewardship regimes performed by the actors involved. They argue that to understand the links between the production, distribution, and use of open data and the types of value it creates, you need to look at which actors are *stewarding* the data and how. This involves examining the practices of, and the dynamics between, the actors and evaluating whether and how their contributions align with the types of value that are desirable in the open development context. Thus, the stewardship approach builds context-driven and practice-based analysis of open data issues that are important to resolve.

Similarly, Rao, Parekh, Traxler, and Ling (chapter 3) examined sharing open educational resources (OERs) and crowdsourcing information to improve public services. They propose a trust model that focuses on how and why relations of trust develop between users, administrators, and sponsors of an open system. Their model enables a comprehensive examination of trust across many different groups of actors and layers of open processes found within an open system. Moreover, they present a series of questions that enable targeted analysis to understand specific relations and practices occurring within the system. Combining multilayered analysis with detailed exposition should help researchers identify trust issues standing in the way of bridging practices and strategic functions of the open system.

In contrast, the desire of Chaudhuri, Srinivasan, and Hoysala (chapter 4) to focus their argument on understanding and overcoming the constraints that individuals and communities face while learning to participate in open practices exemplifies what we see as a fundamental improvement in the way that open development can be investigated. Namely, their approach hones in on links between open practices, situated learning, and identity transformations, providing a series of questions to investigate these links.

The reflections in part I of this volume comment on the benefits and limitations of such pragmatic approaches to open development. Kendall and Dasgupta's (chapter 7) reflection portrays the benefits of Chaudhuri, Srinivasan, and Hoysala's (chapter 4) *learning as participation* framework in their study of a weather information system in West Bengal. Their findings reinforced Chaudhuri, Srinivasan, and Hoysala's (chapter 4) intention to better understand the limits that most marginalized and disempowered people confront in learning substantively from participating in open initiatives. Mungai and Van Belle's (chapter 5) investigation of stewardship regimes and intermediation models within Kenya's Open Data Initiative found Reilly and Alperin's (chapter 2) approach helpful in categorizing the stewardship practices occurring. However, they argued that greater insight into the types of value that generate meaningful use of open data by citizens is still needed. Likewise, Sadoway and Shekhar's (chapter 6) critique of Rao et al.'s (chapter 3) trust model calls for a citizen-centric framing of trust in open development. Their reflection focuses on urban public service provision in the Indian city of Chennai, arguing that the introduction of open initiatives occurs in a particular social and political context with existing levels of trust (and distrust). Thus, the starting point for understanding trust is not with the open system but rather with trust relations in the existing context. These reflections demonstrate how the theoretical frameworks can be used to identify key problems that need to be addressed.

Part II: Coevolutionary Perspectives on Open Development

The pragmatic approaches identified earlier in all likelihood incorporate many of the same ideas and approaches as the coevolutionary perspective. Where they differ is in the latter's firm stance on the necessity of making explicit the connections between actors, institutions (informal and formal), and development discourses and outcomes. This perspective stresses the need not only to root open processes in grounded, local realities but also to transform development discourses, institutional structures, and open practices in a mutually constitutive way. The coevolutionary perspective broadly involves the following:

- Critical intention to transform institutional structures
- Favoring participatory approaches
- Challenging dominant knowledge and production hierarchies

Part II consists of three more chapters. For Dearden, Walton, and Densmore (chapter 8), a major reason why open initiatives fail is because disadvantaged and disempowered people are not involved as *writers* of open development but only as *readers* of it. They are skeptical that divergent outcomes experienced by participants in open processes (such that some people tend to benefit more than others) will be addressed until all participants gain the power to develop *writing relationships* within open processes. Inevitably, this will require confronting dominant knowledge and production hierarchies that can be uncovered through situated activity analysis. In this way, Dearden, Walton, and Densmore provide a solid framing of how they wish to see open development evolve, and their lens incorporates intriguing insights from New Media Literacy to establish how outcomes that are more equitable can be achieved.

A principled approach is also a feature of Zheng and Stahl's (chapter 9) application of the critical capability approach to evaluate open initiatives. They intend to help researchers and practitioners identify and build into projects specific ways to address hegemonic tendencies that respect human diversity, increase democratic discourse, and prioritize human-centered development. By engaging with the evaluation framework, their intention is to awaken actors within their own practices to learn new ways of thinking and acting through openness. All three aspects—actors, open practices, and institutions—are influenced to change as they take part in a coevolution.

Open development is then taken in an innovative direction by Singh, Gurumurthy, and Chami (chapter 10). Starting from the observation that open development conceptualizations based on "process, action, or artifacts, demanding universal access, participation, and collaboration, are unable to adequately account for cases where there is a *justifiably* limited or circumscribed reach of openness," they propose an alternative definition of open development that engages with the institutions of development. Their proposed open institutional design underlines the steps public interest organizations can take to transform institutional structures and development discourse toward increasing transparency, participation, and collaboration.

The reflections in part II of this volume capture the challenges researchers and practitioners confront in applying these frameworks for their intended purposes, as power relations, discourses, and institutional structures are inherently difficult to change. Yet Gamage, Rajapakse, and Galpaya's (chapter 11) reflexive discussion of their own transformation to understand, realistically, what *writing rights* actually mean in their project on sharing of Sri Lankan agriculture information demonstrates optimism for the effectiveness lens provided by the work of Dearden, Walton, and Densmore. However, Moshi and Shao's (chapter 12) application of the critical capabilities approach evaluation framework to Tanzania's Open Data Initiative warranted much less optimism. Their evaluation, while demonstrating the value of Zheng and Stahl's (chapter 9) contribution, suggested a limited impact because of a lack of institutionalization of the initiative within the Tanzanian government. Bentley (chapter 13), in focusing on civil society organizations working in the development sector, likewise suggests that three major problems stand in the way of open institutional design taking root. Singh, Gurumurthy, and Chami's (chapter 10) design may therefore need to inspire a movement such that waves of institutions and publics make open institutional design a priority.

Conclusion

In conclusion, we view the open development theoretical frameworks in this book as an opportunity to support actively engaged scholarship in this area that surpasses the limitations and failures of overly optimistic views based purely on the processes and characteristics of open practices.

Moreover, this volume's contributions resist and confront approaches to open development that serve the interests of the most powerful institutions and actors. Instead, the lenses offer practical and widely applicable insights for organizing open development toward a process of positive social transformation. We envision open development as a worthy and inclusive field of research and practice that supports many diverse cultures and ways of being and doing. We hope to see scholars, practitioners, and policymakers utilize these lenses, especially to improve on the practices and institutionalization of open development in the interests of the world's poorest and most marginalized people.

References

Bates, Jo. 2014. "The Strategic Importance of Information Policy for the Contemporary Neoliberal State: The Case of Open Government Data in the United Kingdom." *Government Information Quarterly* 31 (8): 388–395.

Bauwens, Michel, and Vasilis Kostakis. 2014. "From the Communism of Capital to Capital for the Commons: Towards an Open Co-operativism." *Communication, Capitalism and Critique* 12 (1): 356–361.

Bentley, Caitlin M., and Arul Chib. 2016. "The Impact of Open Development Initiatives in Lower- and Middle Income Countries: A Review of the Literature." *Electronic Journal of Information Systems in Developing Countries* 74 (1): 1–20. https://onlinelibrary.wiley.com/doi/10.1002/j.1681-4835.2016.tb00540.x.

Bentley, Caitlin M., Arul Chib, and Sammia C. Poveda. 2018. "Exploring Capability and Accountability Outcomes of Open Development for the Poor and Marginalized: An Analysis of Select Literature." *Journal of Community Informatics* 13 (3): 98–129. http://ci-journal.org/index.php/ciej/article/view/1423.

Berners-Lee, Tim. 2006. "Linked Data—Design Issues." W3.org. https://www.w3.org/DesignIssues/LinkedData.html.

Buskens, Ineke. 2013. "Open Development Is a Freedom Song: Revealing Intent and Freeing Power." In *Open Development: Networked Innovations in International Development*, edited by Matthew L. Smith and Katherine M. A. Reilly, 327–352. Cambridge, MA: MIT Press; Ottawa: IDRC. https://idl-bnc-idrc.dspacedirect.org/bitstream/handle/10625/52348/IDL-52348.pdf?sequence=1&isAllowed=y.

Cowen, Michael P., and Robert W. Shenton. 2003. *Doctrines of Development*. London: Routledge.

Crush, Jonathan. 1995. *Power of Development*. London: Routledge.

Cyranek, Günther. 2014. "Open Development in Latin America: The Participative Way for Implementing Knowledge Societies." *Instkomm.De*, April 6. http://www.instkomm.de/files/cyranek_opendevelopment.pdf.

Deseriis, Marco. 2017. "Technopopulism: The Emergence of a Discursive Formation." *Communication, Capitalism and Critique* 15 (2): 441–458.

Dryzek, John S. 2000. *Deliberative Democracy and Beyond: Liberals, Critics, Contestations*. Oxford: Oxford University Press.

Escobar, Arturo. 1995. *Encountering Development: The Making and Unmaking of the Third World*. Princeton, NJ: Princeton University Press.

Gray, Jonathan, Jordan Hatcher, Becky Hogge, Simon Parrish, and Rufus Pollock. 2009. *Unlocking the Potential of Aid Information, Version 0.2*. London: Open Knowledge Foundation and AidInfo.

Habermas, Jürgen. 1987. *The Theory of Communicative Action*. Translated by Thomas McCarthy. Boston: Beacon Press.

Heller, Michael. 2008. *The Gridlock Economy: How Too Much Ownership Wrecks Markets, Stops Innovation, and Costs Lives*. New York: Basic Books.

Hodgkinson-Williams, Cheryl. 2015. "Grappling with the Concepts of 'Impact' and 'Openness' in Relation to OER: Current Developments in the ROER4D Project." http://conference.oeconsortium.org/2015/wp-content/uploads/2015/08/Hodgkinson-Williams-2015-final-OEC-Banff-2015.pdf.

Latouche, Serge. 1988. "Contribution à l'histoire du concept du développement [Contribution to the history of the concept of development]." In *Pour une histoire du développemen: État, sociétés et développement* [For a History of Development: State, Societies and Development], edited by Catherine Coquery-Vidrovitch, Daniel Hemery, and Jean Piel, 41–60. Paris: L'Harmattan.

Macmanus, Richard. 2004. "Tim O'Reilly Interview, Part 3: eBooks & Remix Culture." *readwrite*, November 17. https://readwrite.com/2004/11/17/tim_oreilly_int_2/.

Moon, Samuel, and Tim Williamson. 2010. "Greater Aid Transparency: Crucial for Aid Effectiveness." Project Briefing 35, January 2010. Sussex: Overseas Development Institute. https://www.odi.org/sites/odi.org.uk/files/odi-assets/publications-opinion-files/5722.pdf.

Nyamrjiah, Francis B. 2009. "Open Access and Open Knowledge Production Processes: Lessons from Codesria." *African Journal of Information and Communication* 10:67–72.

Open Knowledge International. 2016. "Open Definition 2.1: Open Definition; Defining Open in Open Data, Open Content and Open Knowledge." Opendefinition.org. http://opendefinition.org/od/2.1/en/.

Paviour, Ben. 2016. "Expensive NGO Phone Apps Gather Digital Dust." *Cambodia Daily*, February 25. https://www.cambodiadaily.com/news/expensive-ngo-phone-apps-gather-digital-dust-108937/.

Percy, Tanya, and Jean-Paul Van Belle. 2012. "Exploring the Barriers and Enablers to the Use of Open Educational Resources by University Academics in Africa." In *Open Source Systems: Long-Term Sustainability*, Proceedings of the 8th IFIP WG 2.13 International Conference, OSS 2012, Hammamet, Tunisia, September 10–13, 2012, edited by Imed Hammouda, Björn Lundell, Tommi Mikkonen, and Walt Scacchi, 112–128. Berlin: Springer. https://doi.org/10.1007/978-3-642-33442-9_8.

Pomerantz, Jeffrey, and Robin Peek. 2016. "Fifty Shades of Open." *First Monday* 21 (5): 80–85. https://firstmonday.org/ojs/index.php/fm/article/view/6360/5460.

Powell, Mike, Tim Davies, and Keisha C. Taylor. 2012. "ICT for or against Development? An Introduction to the Ongoing Case of Web 3.0." IKM Working Paper 16. Bonn: IKM Emergent Research Programme; European Association of Development Research and Training Institutes.

Publish What You Fund. 2015. "Campaign for Aid Transparency." *Publish What You Fund*. https://www.youtube.com/watch?v=GJCbHu-cWBw.

Rahnema, Majid, and Victoria Bawtree, eds. 1997. *The Post-development Reader*. London: Zed Books.

Reilly, Katherine M. A., and Matthew L. Smith. 2013. "The Emergence of Open Development in a Network Society." In *Open Development: Networked Innovations in International Development*, edited by Matthew L. Smith and Katherine M. A. Reilly, 15–50. Cambridge, MA: MIT Press; Ottawa: IDRC. https://idl-bnc-idrc.dspacedirect.org/bitstream/handle/10625/52348/IDL-52348.pdf?sequence=1&isAllowed=y.

Restakis, John, Daniel Araya, Maria José Calderon, and Robin Murray. 2015. "ICT, Open Government and Civil Society." *Journal of Peer Production* 7:1–30. http://peerproduction.net/wp-content/uploads/2015/06/ICT-Open-Government-and-Civil-Society.pdf.

Sachs, Jeffrey D., Vijay Modi, Hernan Figueroa, Mariela Machado Fantacchiotti, Kayhan Sanyal, Fahmida Khatun, and Aditi Shah. 2015. *How Information and Communications Technology Can Achieve the Sustainable Development Goals*. New York: The Earth Institute, Colombia University; Ericsson. https://irp-cdn.multiscreensite.com/be6d1d56/files/uploaded/ICTSDG_InterimReport_Web.pdf.

Sadowsky, George, ed. 2012. *Accelerating Development Using the Web: Empowering Poor and Marginalized Populations*. San Francisco: World Wide Web Foundation.

Saif, Umar, Ahsan L. Chudhary, Shakeel Butt, Nabeel F. Butt, and Ghulam Murtaza. 2009. "A Peer-to-Peer Internet for the Developing World." *Information Technologies and International Development* 5 (1): 31–47.

Singh, Parminder J., and Anita Gurumurthy. 2013. "Establishing Public-ness in the Network: New Moorings for Development—a Critique of the Concepts of Openness and Open Development." In *Open Development: Networked Innovations in International Development*, edited by Matthew L. Smith and Katherine M. A. Reilly, 173–196. Cambridge, MA: MIT Press; Ottawa: IDRC. https://idl-bnc-idrc.dspacedirect.org/bitstream/handle/10625/52348/IDL-52348.pdf?sequence=1&isAllowed=y.

Smith, Matthew L., and Katherine M. A. Reilly, eds. 2013. *Open Development: Networked Innovations in International Development*. Cambridge, MA: MIT Press; Ottawa: IDRC. https://idl-bnc-idrc.dspacedirect.org/bitstream/handle/10625/52348/IDL-52348.pdf?sequence=1&isAllowed=y.

Smith, Matthew L., and Ruhiya Seward. 2017. "Openness as Social Praxis." *First Monday* 22 (4). https://firstmonday.org/ojs/index.php/fm/article/view/7073.

Smith, Matthew L., and Ruhiya Kristine Seward, eds. 2020. *Making Open Development Inclusive: Lessons from IDRC Research*. Cambridge, MA: MIT Press; Ottawa: IDRC. https://idl-bnc-idrc.dspacedirect.org/bitstream/handle/10625/59418/IDL-59418.pdf?sequence=2&isAllowed=y.

Transparency International Cambodia. 2017. "Bribespot: Track and Report Bribes in One Click." http://www.ticambodia.org/bribespot-track-report-bribes-one-click/.

Tsing, Anna. 2016. "Consider the Problem of Privatization." In *Feminist Futures: Reimagining Women, Culture and Development*, edited by Kum-Kum Bhavnani, John Foran, Priya A. Kurian, and Debashish Munshi, 45–50. London: Zed Books.

United Nations Global Pulse. 2012. "Big Data for Development: Opportunities and Challenges." White paper. New York: United Nations Global Pulse. http://www.unglobalpulse.org/sites/default/files/BigDataforDevelopment-UNGlobalPulseMay2012.pdf.

Unwin, Tim. 2017. *Reclaiming Information and Communication Technologies for Development*. Oxford: Oxford University Press.

van Reijswoud, Victor, and Arjan de Jager. 2008. *Free and Open Source Software for Development*. Monza: Polimetrica.

Wiley, David. 2014. "The Access Compromise and the 5th R." *Iterating toward Openness* (blog), March 6. https://opencontent.org/blog/archives/3221.

Willis, Katie. 2011. *Theories and Practices of Development*. 2nd ed. London: Routledge.

World Bank Institute. 2014. "Innovation." Wbi.Worldbank.org.

Zambrano, Raúl. 2017. "Blockchain: Unpacking the Disruptive Potential of Blockchain Technology for Human Development." White paper. Ottawa: IDRC. https://idl-bnc-idrc.dspacedirect.org/bitstream/handle/10625/56662/IDL-56662.pdf.

I Pragmatic Approaches to Open Development

2 A Stewardship Approach to Theorizing Open Data for Development

Katherine M. A. Reilly and Juan Pablo Alperin

Introduction

Early open data for development (OD) work was premised on the assumption that IT-enabled open data would decentralize power and enable public engagement by disintermediating knowledge-intensive processes such as education, decision-making, innovation, cultural production, health care, and publishing (Smith, Elder, and Emdon 2011, iv). With this in mind, OD practitioners and researchers have tended to focus their efforts on the distribution of data rather than its production or uptake (Smith and Seward 2017). However, in practice, public engagement in open data has been insufficient (Mutuku and Mahihu 2014), as well as asymmetrical or inequitable (Benjamin et al. 2007) in developing countries, suggesting that much greater attention should be paid to the forces shaping production and uptake of data.

There is increasing recognition that uneven uptake of open resources is more than just a problem of inadequate publicity or capacity. Writing about the *World Development Report 2016: Digital Dividends*, Yochai Benkler (2016) argues that policymakers need to focus their attention on the growing power of platforms to mediate our access to economically productive digital resources. Rather than giving people the skills necessary to access, use, and appropriate open data, says Benkler, governments need to start creating regulations that prevent these platforms from controlling the economic, social, and political opportunities available to citizens through open processes. We are seeing growing recognition of the need to analyze the motivations and agendas of key actors within the open data field. For example, Tyson (2015, n.p.) argues that "data intermediaries will play a critical role in the post-2015 development agenda" because they will determine whether and how to measure the achievement of the UN Sustainable Development

Goals. The suggestion here is that the people who control data about development will be powerfully placed to influence development agendas.

The forces shaping data production, distribution, and use can sometimes feel like foregone conclusions. For example, today it is widely observed that "capitalism has turned to data as one way to maintain economic growth and vitality in the face of a sluggish production sector," as well as that "the platform has emerged as a new business model capable of extracting and controlling immense amounts of data, and with this shift we have seen the rise of large monopolistic firms" (Srnicek 2017, 6). Forces such as these, which can feel inevitable, are an external threat to open data even as they influence the way open data is internally managed.

In the face of these forces, this chapter argues that if we want to know how and why people engage with *open* data, we need to uncover the actors and governance regimes shaping those engagements. In other words, we need to study how open data is *stewarded* by the actors who manage it.

We are using the word stewardship deliberately to emphasize that *all* data are a resource and that these resources can be managed or regulated in a variety of ways. This serves to remind us that patterns of intermediation or governance of data are not standard operating procedures that have arisen historically. Patterns of data governance and data intermediation are a *choice* that actors and communities make about the stewardship of their knowledge commons. These choices are shaped by the values that actors attempt to extract from data resources, and the forces governing these choices should be a focus of our analysis. Patterns of intermediation and governance reflect the collective struggle to extract different types of value from data resources, and they will determine who benefits from data resources and how. At times, actors may come together in common regimes of data governance, but they may also pull on data resources in unintended, contradictory, or conflicting ways. This applies to open data just as it does to any other kind of data, so we can use stewardship as a flashlight to focus our attention on these processes, their implications for the knowledge commons, and outcomes for development.

In what follows, we explore the concept of stewardship as a way of thinking about engagement with open data. We explain the idea of stewardship as it relates to open data and then discuss stewardship in relation to governance regimes and actors. Finally, we consider the literature on intermediation as an avenue through which to explore different models of stewardship of open data. We offer four models of data intermediation and point out the different combinations of values that shape choices about data stewardship in each case. In teasing out differences in regimes and actors between the four models, we demonstrate how this approach can be applied.

Stewardship of Open Data

In general terms, stewardship is the management, safeguarding, and enhancement of goods that belong to others. Data stewardship is a specialized field of data management that ensures the quality of data by creating systems that are in compliance with regulatory obligations (Plotkin 2013). This definition works well to describe the individuals who manage data assets and systems within relatively closed systems, but when it comes to *open* data, we need to push this definition a bit further.

This is because open data is publicly available, so it can be reused and redistributed.[1] In the world of open data, there is a tendency to focus on the qualities of the *data*—to ask whether it is managed in such a way that it can be openly accessed—but, in fact, what matters more are the typical information and communications technology (ICT) enabled social processes that allow replicable data to be reused over and over again in a variety of ways. This is partially captured by the idea of prosumption. Prosumption is a cocreation arrangement in which the *consumers* of a resource also do the work of producing it. This means that open data is only really *open* to the extent that it is taken up in open processes, such as peer production, crowdsourcing, sharing, collaboration, or reuse, revision, or remixing (Smith 2014). So, in order for it to be considered *open*, not only the data but also the social networks surrounding that data need to be addressed in models of data stewardship.

This being the case, stewardship of open knowledge production processes is about the management, safeguarding, and enhancement of the knowledge commons. Stewardship situates data production, distribution, and uptake in the wider historical, institutional, and political contexts that determine how the burdens and benefits of data are shared. It is with this concept in mind that Block defines stewardship as "the choice to preside over the orderly distribution of power" (Block 2013, xxiv). This means that systems of data stewardship will be the object of political struggle as different interest groups seek to advance the vision of data management that best suits their ends. We also see people asserting their vision for the governance of open data within the knowledge commons. For example, Block argues that stewardship of open data should prioritize the common good as well as private return (Block 2013; see also Wagner 2013).

This leads us to a larger discussion about how data *ought* to be stewarded. A major assumption of OD is that openness will result in more resilient social processes and a more equitable distribution of the use, social, or exchange value of knowledge resources. However, the value of open data is based on stakeholder interests that do not always align with OD objectives. *Use value* is the immediate utility of open data in a given context. For example, someone might decide that a data set is useful in answering a specific question. *Exchange value* is the monetary value of open data, which includes both sunk costs and potential profit. For example, people may choose not to engage with open data because they see that the cost of manipulating data to make it useful for a particular purpose is too high. However, if they do invest time and resources in making the data actionable, they may seek to recover their costs and/or make a profit from their efforts.[2] Finally, *social value* is the wider social benefit that results from the creation of either use value or exchange value. When an intermediary decides to improve the quality of data, perhaps because they need it for their own purposes, but leave the data *out there* for others to use as well, then a socialized value has been created.

Achieving a positive result for development depends on the type of stewardship model that is put in place—in other words, the types of incentive systems, conventions, cultural understandings, institutional mechanisms, and moral contracts put in place to steward open knowledge production processes. These stewardship arrangements will shape the distribution of responsibilities for production, maintenance, and use of data resources, as well as flows of value that emerge from them. We can better understand whether and how *open* data initiatives *make a difference* by studying arrangements to steward knowledge and the effects of these arrangements on the distribution of value from those goods. In doing so, we can arrive at a better understanding of the factors that link openness to social change outcomes.

This is a challenging proposition because openness itself is shifting our understanding of resources (Spence and Smith, forthcoming). It is sometimes assumed that open resources are necessarily a public good, given that openness initiatives can arise in situations where information is nonexcludable and also sometimes nonrivalrous (Tennison 2015). However, this is not entirely accurate because even assuming ubiquitous, cheap, and open computer networks, information is subject to controls (e.g., state secrecy), capture (e.g., intellectual property), and investment costs (production and maintenance). Also, even though information may be nonrivalrous, user time and audience attention most certainly are. The whole point of the public goods discussion is to justify public sector intervention in the marketplace in cases of *market failure*. When it comes to stewardship of open data, however, it is unclear whether or how the public sector should intervene.

Stewardship therefore demands careful consideration of how—through what arrangements—open resources can best be provided and how best to maximize the quality, sustainability, buy-in, and uptake of those resources. For example, peer production licenses, as discussed by Bauwens (2013), aim to ensure that public goods are used in ways that sustain the knowledge commons while also reallocating resources from the private sector to the maintenance of that commons. In this case, the debate over public versus private provision takes a back seat to strategies that socialize the use value and social value generated through joint production of an informational good (Bollier 2014; Meng and Wu 2013). The question then is how to ensure that actors extract and share the use, social, and exchange values of that good in ways that also sustain and enhance the knowledge commons.

Stewardship also challenges us to think about how—through what arrangements—different actors become *engaged* in openness initiatives. Public engagement is different from public participation (Bovaird 2007; Bovaird and Loeffler 2012). Participation implies that targeted invitations are extended to people who can contribute in preconceived ways. In this sense, participation tends to be *transactional* in nature. Engagement, however, implies motivated and reflexive contributions to a jointly produced and therefore evolutionary context. It recognizes the dissolution of the boundary between user and producer in the management of the resource, as well as the shifting balance of costs and benefits between different user groups, which might include a wide range of actors (Gençer and Oba 2011). As a result, engagement is said to be *transformative* in nature. In openness initiatives, stewardship should contemplate active public engagement, both at the level of governance decisions and at the level of data production and management, to maintain the openness of the initiative. But whether and how stewardship of open data systems is managed will depend on how different actors view the value in supporting open systems, a problem that we turn to in the next section.

Locating Stewardship: Regimes and Actors

To study patterns of stewardship as described earlier, it is necessary to identify the governance regimes and actors at work within a particular setting. That setting may be very local, involving a small, closely knit community of actors, or it may be national or even global, encompassing a wide range of disparate actors. Governance regimes establish the policies and approaches that enable and constrain data stewardship arrangements.

Approaches to governance will vary widely depending on the field of analysis and the interests of actors. For example, Rosenbaum argues that within the American health care system, the major concern shaping governance of health data is the correct balance of privacy and access. Rosenbaum (2010, 1443) comments that "an intense struggle over health information access has been a hallmark of the healthcare system for decades." In this case, health data stewardship focuses on "the broad policies for access, management, and permissible uses of data; identifies the methods and procedures necessary to the stewardship process; and establishes the qualifications of those who would use the data and the conditions under which data access can be granted" (Rosenbaum 2010, 1445). This example helps us see how decisions about how to steward data shape the work of data intermediaries and also the ability of other actors to engage with health data. In this case, the social value of privacy and the use value of data access tend to outweigh the exchange value that could be extracted from health data in the determination of governance regimes around data management. In other sectors and places, the central issues driving debates over governance will differ, as will the balance of values held by key stakeholders, so stewardship regimes will be different.

Much of the literature on data stewardship focuses on enterprise-level data management, and thus it emphasizes the need to align a single firm's stewardship principles and practices with its goals and internal operational culture. Much of this literature starts by extolling the virtues of good data management for planning, decision-making, performance, and the like. It then goes on to provide guidance on how to organize data so that it supports the main agendas of the organization (see, for example, Ballard et al. 2013). The value of this advice does not change for individual organizations in a world with open data. Private sector organizations will continue to organize their data management practices to match their specific business practices and organizational goals. This means that decisions about how to manage data will hinge on business models and may prioritize different styles of data stewardship, which may use either open or proprietary data.

We can extend this discussion beyond the private sector to government, service, and civil society organizations. These, too, will attempt to align internal principles for data governance to organizational models and objectives. Thus, we can expect nongovernment organizations to emphasize the importance of social agendas, such as diversity, in their data governance mandate, while government offices are likely to emphasize public service goals, such as timeliness, in their approaches to data governance. The different ways these organizations steward their data will align with the particular interests and goals of the organization.

This is an important point to contemplate. There is a tendency to assume that the impacts of open data on development can be studied across a whole society, that opening up government data or crowdsourcing societal data will have a generalized benefit. But when we take a stewardship approach to thinking about data, we quickly discover that data management practices differ widely across a community of actors, depending on the values and agendas that they bring to the data. Each organization will see a different use value in a particular data set. As a result, they will invest in the data in different ways. Sometimes this will literally mean transforming the data into usable material, which means that the resource carries an *added value*. This has implications for the exchange value of the data and the desirability, utility, or even *feasibility* of sharing the data back into the community. The need or desire to create exchange value may then contradict efforts to create social value through open data.

Once we accept this fact, it becomes quite difficult to imagine a data governance regime at, say, the national level because different actors have such different needs and processes of uptake with regard to data. In this sense, open data is most easily understood as a *resource* that can be extracted by different actors in different ways to address their various mandates. The only way open data *systems* can become more fully presumptive is if they are organized around a common agenda, such as software production or communal knowledge mobilization for things like mapping of resources or observation of the natural world. Otherwise, what is more likely to occur is disjuncture or competition in open data regimes.

What we learn from this discussion is that it is necessary to study the governance regimes that emerge around specific instances of open data if we are to understand stewardship practices in particular contexts. To do this, it is important to identify the actors involved in data governance and their interests.

There have been attempts to model the actors involved in making decisions about open data. For example, a typical model differentiates between data producers, data intermediaries and their platforms, and data users (van Schalkwyk et al. 2015). This model is based on the idea of data devolution. One of its blind spots is its tendency to assume that all actors within a particular category will approach data governance in the same way. For example, it assumes that all intermediaries work to organize data such that end users can better access it. Another blind spot in this model is the assumption that data producers, intermediaries, and users exist in a linear relationship. This can be the case, for example, when governments open up data that intermediaries organize to make useful for citizens. But in other cases, such as with crowdsourcing initiatives, end users may devise a platform that collects and organizes the data they require to achieve an end.

A stakeholder approach can help to resolve this difficulty. For example, the open data life cycle model by van den Broek, van Veenstra, and Folmer (2014) identifies top managers, information managers, legal advisers, community managers, and data owners as key stakeholders involved in the process of opening up data. This is helpful, especially since it differentiates between different types of data managers, who will have different categories of concerns in relation to data management, ranging from technical to strategic. In other words, this model helps us identify the interests and goals shaping both the whole open data initiative and also those of individual stakeholders, which may be in tension with each other. But, in turn, this model lacks an analysis of the relationships between these actors.

Gonzalez-Zapata and Heeks (2015) offer a solution that organizes stakeholders according to their power to influence the implementation of open data programs and their level of interest in doing so. Stakeholders in this example, which referred to open government data, included politicians, public officials, international organizations, ICT providers, donors, academics, and civil society actors. However, it is easy to see how the same model might be extended to other cases.

Taking these three models together, it is possible to identify how different actors influence the rise of governance regimes around open data within a community of actors and how different actors organize their own activities in response to this emerging set of practices and principles. In what follows, we review the literature on intermediation and identify different patterns of data governance implied by different assumptions about the actors who work with open data. In each case, we identify the values driving governance of data and discuss how each of these models can help us understand patterns of stewardship that may arise in real-world open data interventions. We refer to these patterns alternatively as schools or models to suggest a unified set of assumptions and goals.

Illustrating the Stewardship Approach: Four Models of Stewardship

The focus of this literature review is on what we might call *primary* or *first-order* information intermediaries. According to van Schalkwyk et al. (2015, 6), "An open data intermediary is an agent: (1) positioned at some point in a data supply chain that incorporates an *open* dataset, (2) positioned between two agents in the supply chain, and also (3) facilitates the use of open data that *may* otherwise not have been the case." These are distinct from other types of intermediaries that either produce data or are the subject of data. For example, open data about aid flows can be used to track the effectiveness of intermediary organizations in international development, and these intermediaries themselves produce data. However, these organizations are not themselves infomediaries, even though they do have a role in producing data.[3]

Data intermediaries are also distinct from the larger *secondary* forces that mediate information flows. We recognize that policy, the media, culture, institutions, language, and technology all work to *mediate* (or create the context for) the knowledge society, but these are not the focus of this chapter. For example, where does the Internet end and platforms begin? ICTs are a particularly difficult case, given that the policies and business models that structure ISPs, search engines, and social networking platforms often directly influence the work of information or data intermediaries. Furthermore, open data presumes access to technologies and the ability to use them, and the nature of that access or use significantly shapes the flow of open resources. All the same, our focus is on technology actors or technology effects rather than on the technology itself.

Finally, this literature review takes a broad approach to thinking about knowledge and therefore addresses data, information, *and* knowledge, at times somewhat interchangeably. It would certainly be possible to produce a literature review that focused only on intermediation of data. However, it should be noted that (1) data is already prefigured through its collection processes; (2) open data is more prevalent in some areas of OD work, while open information dominates in others; and (3) many intermediaries work across different forms of data (e.g., budget numbers) and information (e.g., WikiLeaks files). Indeed, intermediaries may bring these different kinds of materials into conversation with each other in the stewardship work that they do.

Based on this review, we identified four patterns of data stewardship, which are summarized in table 2.1 and discussed here.

Table 2.1
Schools of thought about intermediation

School of thought	Central assumption	Groups involved	Central aim	Tools and methods
Arterial school	Intervention is required to assure flows from producers or holders to users	Infomediaries	Information flow from holders to users	Public access computing (PAC); open analytics; training; education and awareness campaigns
Ecosystems school	Open data comes from many places and is used in many ways, so you need a complex array of innovators to extract value	Civic start-ups; open data services; datamediaries; data wranglers	Innovation; value added; more broadly, problem solving, economic growth, and institutional development	Aggregation; hackathons; data jams; crowdsourcing; ledgers; linked data
Bridging school	Raw materials difficult to understand; mediators work to "make data actionable"	Journalists; advocates; programmers; technical or science communicators	Bridging social values with foreign, scientific, or bureaucratic logic	Translation; facilitation; localization
Communities of practice	Strong norms are required to mediate the management of open knowledge for learning, innovation, and other uses	Organizations; networks; epistemic communities	Facilitate productive collaboration; maintain data commons	Information architecture; norms of governance

Arterial School

The arterial school recognizes that even when data or information is made freely available on the Internet, people often face obstacles when accessing it—that there are blockages in the informational arteries that reach out into society. Commentators often point out that just opening up the data is not enough to ensure awareness, use, or engagement. Intermediaries or *infomediaries* are prescribed as a means of overcoming barriers. As a result, the main governance criteria used for data stewardship in the arterial school are the democratization of data and the decision-making processes around that data. This is sometimes complemented by a capabilities approach for accessing the data, making decisions about it, and mobilizing it.

Originally, this discussion focused on supporting *access, use,* and *appropriation* of ICTs through public access computing (PAC) at libraries, cybercafés, and telecenters (Gomez, Fawcett, and Turner 2012; Sein and Furuholt 2012) or the work of community organizations (Beck, Madon, and Sahay 2004). This work has also been explored within specific domains of stewardship. Al-Sobhi, Weerakkody, and Kamal (2010), for example, research the intermediary organizations that facilitate coordination between public services and users. As they point out, "The intermediary provides a trusted information channel gateway and also provides help and support, which may have an impact on citizens' usage toward e-government services" (Al-Sobhi, Weerakkody, and Kamal 2010, 2; see also Sein 2011).

More recently, as attention has turned to open data, the emphasis has shifted toward platforms and tools that help people make sense of open information (such as data visualization tools). Gurstein (2011, n.p.) argues, for example, that in order to overcome the data divide, it is necessary to ensure that "those for whom access is being provided are in a position to actually make use of the now available access (to the Internet or to data) *in ways that are meaningful and beneficial for them*" (emphasis added). Gurstein (2011, n.p.) also expresses concern that open data "empowers those with access to the basic infrastructure and the background knowledge and skills to make use of the data for specific ends" and that it may also "further empower and enrich the already empowered and the well provided for rather than those most in need of the benefits of such new developments." He advocates an effective use approach to open data that would use training programs to ensure that "opportunities and resources for translating this open data into useful outcomes would be available (and adapted) for the widest possible range of users" (Gurstein 2011, n.p.).

Janssen, Charalabidis, and Zuiderwijk (2012, 264) similarly argue that it "cannot be expected that the public has the same amount of knowledge and capabilities as researchers do. Lowering the knowledge level required for use is key to large-scale dissemination." Tools such as data visualization exist to lower barriers. They require, however, "that current efforts take the user's perspective into account and monitor the need, ultimately helping users and lowering the threshold to using open data" (Janssen, Charalabidis, and Zuiderwijk 2012, 265).

However, even if people do have access to data, they may not have clarity about how to mobilize it, and when it comes to organizing that data in strategic ways, expertise may be required. Baack (2015, 6) argues that "Even though the idea behind the democratization of information is to *potentially* allow *everybody* to interpret raw data, activists are well aware that the average citizen does not have the time and expert knowledge to do so." With this in mind, Baack (2015, 6) calls for *empowering intermediaries* that are "*data-driven*, which means that they should be able to handle large and complex datasets to make them accessible to others, … *open*, which means that they should make the data from which they generate stories or build applications available to their audiences, … [and] *engaging*, which means that they should actively involve citizens in public issues."

This school is sometimes referred to as the *one-way street* model of intermediation (Pollock 2011) because much of the literature focuses on ensuring that marginalized users gain access to information that comes from centralized information sources. Additionally, this school is often more concerned with making data flow outward from centers of power than with creating information feedback loops. So, for example, the discussion often revolves around ensuring that citizens are able to access and make sense of government information, but less attention is paid to how the data work of citizens can flow back into decision-making processes. But then we must recognize that different intermediaries and/or different communities will likely arrive at very different visions of how data flows should be organized and what purposes they will serve.

In this sense, the arterial school is focused on the raw potential use value of data but pays little attention to the extra value that must be added before the data can be deployed to achieve a particular end. Nevertheless, without widespread access to open data, it is difficult to ensure that use, social, or exchange value is generated and distributed throughout society. As a result, the arterial school advocates strongly for a knowledge stewardship regime that favors open distribution of knowledge resources in society. This sets the stage for a somewhat naive tendency to overlook the investment that goes into data, and an assumption that use or exchange value extracted from data is at worst nonrivalrous and at best complementary. There is also little discussion of how the use value or social value of the data will manifest. A stewardship approach can precisely highlight the dynamics that animate the work of arterial intermediaries and how their interventions serve to prioritize particular kinds of values in data.

Ecosystems School

The ecosystems school observes that in complex institutional relationships, as between a government and its stakeholders, data is generated by many different information systems that are attached to a wide variety of different social processes. The goal of the ecosystem school is to create a stewardship regime that ensures the production of *quality* data or information that will produce value. This requires careful analysis of a variety of intermediaries and the many ways in which they add value within the ecosystem, as well as the policies and systems that support those intermediaries (Harrison, Pardo, and Cook 2012; Heimstädt, Saunderson, and Heath 2014). As Harrison, Pardo, and Cook (2012, 910) point out, in a data ecosystem, "leaders must engage in a kind of strategic ecosystem thinking" aimed at managing intentionality, value creation, and sustainability. Harrison, Pardo, and Cook (2012, 912) write that "Ultimately, the value of open data rests on *whether or not it enables us to solve problems and meet important needs of individuals, communities, or society writ large*" (emphasis added). This approach raises important questions about what constitutes *value creation* in open data stewardship.

Early works on information intermediaries arose in the industrial management literature. Rose (1999, 76), for example, notes that "Information intermediaries are economic agents supporting the production, exchange, and utilization of information in order to increase the value of the information for its end-user or to reduce the costs of information acquisition. ... The aim to make profit is the origin of their activities. The information processing activities of information intermediaries can generate an informational surplus or added value."

This line of thinking became significant in the aftermath of the 2008 financial crisis as governments, particularly in the United Kingdom, sought new foundations for economic growth. In this context, open government data came to be seen as a possible stimulant for information-age industrial productivity. Thus, for example, Janowski, Holm, and Estevez (2012, 2), in a briefing note titled *Open Growth: Stimulating Demand for Open Data in the UK*, suggest that intermediaries are the supporting industries, such as data management and storage companies, platform and software providers, crowdsourcing hosts, and advisory services *plus* application developers and businesses that occupy the space between open data suppliers and final consumers. The latter take open data and enrich and add services to it so that it can be used by governments, businesses, and individuals. This benefits the wider economy by providing economic growth, increased innovation, and efficiency savings. In other words, leverage of open data is driven by the pursuit of exchange value, with the hope that this will create wider social value.

This thinking has shaped the literature about partnerships between governments and the business community. For example, Sorrentino and Niehaves (2010, 1) argue that, in the future, "eGovernment will be increasingly built on public-private partnerships and will introduce new intermediaries to the public service delivery chain and democratic processes." With this in mind, they address e-government as an open system in which "rational or efficiency-based forces are not the only drivers at work" (Sorrentino and Niehaves 2010, 2). Sorrentino and Niehaves (2010, 3) observe that in some studies the focus is on providing access to public services but that in other cases intermediaries are associated with "the ability to process, generate, and (re)combine data and information" with the realization of a specific social value in mind. This kind of thinking gives rise to studies on different business models for open data intermediation (Janssen and Zuiderwijk 2014).

This literature has influenced work on data intermediation in developing country contexts (e.g., Chattapadhyay 2014). For example, see van Schalkwyk et al. (2015, 4) and also the observation made by Magalhaes, Roseira, and Strover (2013) that "the ICT ecosystem is driven by innovation (i.e., the injection of new knowledge into the ecosystem). Firms compete and co-operate symbiotically, and the interaction between firms and consumers (that is, between knowledge creators and knowledge consumers) generates new knowledge which leads to innovation in the ecosystem. It is the pursuit of innovation that keeps the ICT ecosystem in motion."

These works stand in contrast to pieces like MacKinnon et al.'s (2014) UNESCO study on the role of intermediaries in fostering freedom of expression. In their work, they found that "Internet intermediaries are heavily influenced by the legal and policy environments of states" and "many state policies, laws, and regulations are—to varying degrees—poorly aligned with the duty to promote and protect intermediaries' respect for freedom of expression" (MacKinnon et al. 2014, 10 and 180).

These works raise questions about the kinds of values that drive stewardship of data within open data ecosystems. At times, this literature seems to suggest that a healthy open data ecosystem is necessary to support solutions to complex social problems, whereas at other times it expresses a need for industrial policy to ensure the viability, innovativeness, and economic productivity of open data ecosystems. Innovation is a common theme within this literature; however, it is often unclear whether the literature is referring to new forms of intermediation, new approaches to social entrepreneurship, or the creation of new tech clusters. As a result, governance of open data can become politicized and prone to capture by groups with different interests, and, in particular, this model seems to favor the creation of exchange value, which may contradict the pursuit of specific kinds of use value or the realization of wider social values. A stewardship approach can help us identify the stakeholders related to these contradictions and the competition or policy struggles that result, drawing attention to the historical production of the knowledge sphere in general and of open data systems specifically.

Bridging School

The bridging school recognizes that it can be difficult for people to make sense of open data. Mediators may be required to help make data *actionable* or reconcile different types of information before it can have use, social, or exchange value. Whereas the arterial school gives people tools to help them arrive at their own conclusions, in this case mediators help to create consonance between disparate pieces of information, as when they work to bridge foreign, scientific, or bureaucratic logic; historical context; and specific social values. Bridging activities might include translation of information between languages or formats or facilitation of conversations between data experts and concerned citizens. Bridging also encompasses *localization* of open resources within specific cultural contexts, something that teachers who work with open educational resources must often do (Li, Nesbit, and Richards 2006).

These mediators bring a unique set of skills to the stewardship of open data. Their work can often tend more toward the consolidation of meaning than toward the facilitation of decentralized power in meaning-making processes. As a result, this school may be controversial among proponents of decentralization but welcomed by those who seek to create a multiorganizational ecosystem. Nonetheless, it is important to recognize that actors such as journalists, activists, and science communicators facilitate processes of meaning-making (Grabill and Simmons 1998; Tauberer 2014). As Tauberer points out (2014, n.p.), "The iconic mediators of the 20th century were the radio and television anchors.... Today's mediators include traditional journalists, but also issue advocates, organizers, and app builders—not just programmers, but statisticians, designers, and entrepreneurs—who make information actionable." Supporters of decentralization may recognize that this is necessary; however, they would likely recommend that bridging actors reveal their sources and processes so that citizens are able to make their own assessments about the quality of the analysis. While this is certainly a good idea, the situation is more complex. Bridging actors may prefigure the production or analysis of data by setting the social, political, or economic agenda, as when a teacher sets the agenda for learning in a course. The bridging school reminds us that social realities are always constructed and that mediators are active in stewarding these knowledge-intensive processes where open data is concerned. The bridging school also demonstrates that the use value of data is not always immediately clear and that it is often necessary to invest heavily in that data before its social value or exchange value becomes clear. In other words, the value of data arises not only from the data itself or from the work that is put into cleaning up databases but also from the context that is created for the application of the data. The stewardship approach directs us to look at the power of bridging actors to influence the implementation and uptake of open data programs through the discursive context that they establish for them.

Communities of Practice

Finally, the communities of practice literature addresses situations in which intermediaries steward common pool resources (rather than public goods). This school takes its inspiration from the work of institutional economist Elinor Ostrom (Hess and Ostrom 2006). In some situations, there is limited incentive for people to share data or information, though the benefits of sharing would be high. For example, data associated with research is often tightly controlled because it requires a great deal of expertise and specialized infrastructure to produce, it is difficult to secure the information once leaked, and everyone races to be the first to extract benefits from the data (Borgman 2015, 73). However, it is widely recognized that data sharing can create efficiencies in the research process and also generates collaborations that could increase the rate of innovation resulting from research processes.

An excellent example of a communities of practice approach to stewardship of open data is offered by Map Kibera, a crowdsourced mapping project in a slum neighborhood of Nairobi, Kenya. Berdou (2011), in her analysis of this project, explores the challenges involved in governing the map as an information commons that requires the active input of community members and that also aims to produce benefits for that same community. Governance of common pool resources can be tricky. The goal is to facilitate productive collaboration on the basis of quality, collaboratively produced data, but potential participants may not feel motivated to contribute or may lack trust in the initiative, as is noted by Borgman (2015, 73): "The success of the knowledge commons depends on the ability to limit enclosure, to make exclusion difficult, and to sustain effective governance models. Libraries, archives, data repositories, and other shared-information resources are under continuous threat of free riders, enclosure, and sustainability."

What we learn from the community of practice perspective is that it may be easier to realize effective governance arrangements for open data when there is a cohesive community of actors who share a knowledge production goal and see immediate benefits from sharing their resources. This implies limiting the scope of the openness of data. In other words, social value for some may come at the cost of exchange value or use value for others. Similarly, prioritizing use value or exchange value for a few people may come at the cost of maintaining social value for a large group of actors. From a stewardship point of view, this is not surprising because openness will be managed and operationalized according to a governance logic that suits a particular end. Rather than situating openness as an ideal to be achieved and focusing our research on cases that qualify as *open*, stewardship tells us that we should study the data governance regime that is emerging as a result of the value that people see in the data. This will ultimately help us to understand when, whether, and how data serves development ends.

Discussion and Conclusions

From this review, we have learned that there are different ways of thinking about stewardship of open data and that each approach fundamentally shapes patterns of rights and responsibilities for different actors. Patterns of stewardship shape who does the work and who gets the rewards from digital production, and therefore will determine the impact of open development initiatives for poor or marginalized populations.

The stewardship approach is borne out in our literature review on intermediation, which facilitated revealing some general trends. We note that the type of value created varies widely in each model, as can the goal of stewardship in general. In communities of practice, for example, the value of data is collectively realized, but commercial measures of value (exchange value) tend to prevail in the ecosystem model. Meanwhile, while in theory each model could be used to achieve any number of different ends, the immediate purpose of each model is different (i.e., reconciliation of different ways of knowing in the case of bridging, information flow in the case of the arterial model, and open access in the decentralization school).

It is important to note that these are merely theoretical models. In the real world, data intermediaries and the other stakeholders they come into contact with are in a daily struggle to figure out how to extract value from open resources. They will likely find themselves experimenting with different approaches, and therefore the same organization or individual may engage in a variety of types of intermediation that cut across the four models. These actors may find themselves attempting to extract different types of value from data (use, social, exchange), depending on their needs at a given time. Having said this, actors may also become invested in a particular approach to thinking about or engaging with open data, and this may cause them to enter into conflict with other actors who take a different view on stewardship of data resources.

The effects of each type of intermediation are still not well understood. OD work has not yet widely embraced the role of knowledge stewardship, nor has OD studied knowledge stewardship in depth. Since open data and open data intermediation are relatively new activities, it is likely that patterns of data intermediation and governance are emergent and that actors who engage in this type of work are actively struggling to generate new data stewardship regimes in their sectors, communities, or even nations. It is particularly interesting to contemplate the types of struggles that these actors face as they make decisions about how to use and work with open data, and the factors that drive them to pick one approach over another or to value open data resources in each of the different ways presented.

Ultimately, we believe that a stewardship approach to thinking about data governance can better help us understand the implications of OD for processes of social change. As Livingstone (2009, 8) notes, the larger question is "whether the mediation of micro processes of social interaction influences macro-historical shifts in institutional relations of power." We can achieve this result by examining the business models and policies that shape stewardship, the networks of stakeholders involved in these processes, and the discourses that influence our thinking about intermediation and stewardship of open information. The results of this work can support policy decisions about the management of informational resources. When thinking about the Internet for development today, Benkler (2016, n.p.) observes, "It's not about skills and productivity, it's about power." Understanding how that power is best stewarded is key to opening up future development potential.

Notes

1. See *Open Data Handbook* (n.d.), http://opendatahandbook.org/guide/en/what-is-open-data/.

2. In the past, these *profits* have come in the form of reputation (Weber 2005). In the Ecosystem section, we explain how the 2008 financial crisis put enormous pressure on open data to demonstrate economic returns.

3. This can be a difficult distinction to maintain. See Davies (2016) for an example that does not maintain this distinction.

References

Al-Sobhi, Faris, Vishanth Weerakkody, and Muhammad M. Kamal. 2010. "An Exploratory Study on the Role of Intermediaries in Delivering Public Services in Madinah City: Case of Saudi Arabia." *Transforming Government: People, Process and Policy* 4 (1): 14–36.

Baack, Stefan. 2015. "Datafication and Empowerment: How the Open Data Movement Re-articulates Notions of Democracy, Participation, and Journalism." *Big Data and Society* 2 (2): 1–11. https://doi.org/10.1177/2053951715594634.

Ballard, Chuck, Trey Andersen, Lawrence Dubov, Alex Eastman, Jay Limburn, and Umasuthan Ramakrishnan. 2013. *Aligning MDM and BPM for Master Data Governance, Stewardship, and Enterprise Processes.* Armonk, NY: International Business Machines Corporation.

Bauwens, Michel. 2013. "Can the Commons Compete with the For-Profit Economy?" *OuiShare Magazine,* July 8. https://www.ouishare.net/article/can-the-commons-compete-with-the-for-profit-economy.

Beck, Eevi, Shirin Madon, and Sundeep Sahay. 2004. "On the Margins of the 'Information Society': A Comparative Study of Mediation." *The Information Society* 20 (4): 279–290. http://heim.ifi.uio.no/~sundeeps/publications/OnTheMarginsOf.pdf.

Benjamin, Solomon, R. Bhuvaneswari, P. Rajan, and S. Manjunatha. 2007. "Bhoomi: 'E-Governance', or, an Anti-politics Machine Necessary to Globalize Bangalore?" CASUM-M Working Paper. Bangalore: Collaborative for the Advancement of Studies in Urbanism through Mixed-Media. https://casumm.files.wordpress.com/2008/09/bhoomi-e-governance.pdf.

Benkler, Yochai. 2016. "What the World Bank Report on Tech-Related Income Inequality Is Missing." *The Guardian,* January 15. http://www.theguardian.com/technology/2016/jan/15/world-bank-income-inequality-report-silicon-valley.

Berdou, Evangelia. 2011. "Mediating Voices and Communicating Realities: Using Information Crowdsourcing Tools, Open Data Initiatives and Digital Media to Support and Protect the Vulnerable and Marginalized." Final research report. Brighton: Institute of Development Studies. https://assets.publishing.service.gov.uk/media/57a08ab7e5274a31e000072e/IDS_MediatingVoices_FinalReport.pdf.

Block, Peter. 2013. *Stewardship: Choosing Service Over Self-Interest.* 2nd ed. Oakland: Berrett-Koehler.

Bollier, D. 2014. "Bauwens: Use a Peer Production License to Foster 'Open Cooperativism.'" *David Bollier: News and Perspectives on the Commons* (blog), March 22. http://bollier.org/blog/bauwens-use-peer-production-license-foster-%E2%80%9Copen-cooperativism%E2%80%9D.

Borgman, Christine L. 2015. *Big Data, Little Data, No Data: Scholarship in the Networked World.* Cambridge, MA: MIT Press.

Bovaird, Tony. 2007. "Beyond Engagement and Participation: User and Community Coproduction of Public Services." *Public Administration Review* 67 (5): 846–860.

Bovaird, Tony, and Elke Loeffler. 2012. "From Engagement to Co-production: The Contribution of Users and Communities to Outcomes and Public Value." *Voluntas: International Journal of Voluntary and Nonprofit Organizations* 23 (4): 1119–1138.

Chattapadhyay, Sumandro. 2014. *Opening Government Data through Mediation: Exploring the Roles, Practices and Strategies of Data Intermediary Organisations in India.*

Delhi: Open Data Research Network. http://www.opendataresearch.org/content/2015/757/opening-government-data-through-mediation-exploring-roles-practices-and-strategies.

Davies, Tim. 2016. "Following the Money: Preliminary Remarks on IATI Traceability." *Tim's Blog: Working for Social Change; Exploring the Details; Generally Quite Nuanced* (blog), January 15. http://www.timdavies.org.uk/2016/01/15/following-the-money-preliminary-remarks-on-iati-traceability/.

Gençer, Mehmet, and Beyzer Oba. 2011. "Organising the Digital Commons: A Case Study on Engagement Strategies in Open Source." *Technology Analysis & Strategic Management* 23 (9): 969–982.

Gomez, Ricardo, Phil Fawcett, and Joel Turner. 2012. "Lending a Visible Hand: An Analysis of Infomediary Behavior in Colombian Public Access Computing Venues." *Information Development* 28 (2): 117–131.

Gonzalez-Zapata, Felipe, and Richard Heeks. 2015. "The Multiple Meanings of Open Government Data: Understanding Different Stakeholders and Their Perspectives." *Government Information Quarterly* 32 (4): 441–452.

Grabill, Jeffrey T., and W. Michele Simmons. 1998. "Toward a Critical Rhetoric of Risk Communication: Producing Citizens and the Role of Technical Communicators." *Technical Communication Quarterly* 7 (4): 415–441.

Gurstein, Michael B. 2011. "Open Data: Empowering the Empowered or Effective Data Use for Everyone?" *First Monday* 16 (2), February 7. http://journals.uic.edu/ojs/index.php/fm/article/view/3316.

Harrison, Teresa M., Theresa A. Pardo, and Meghan Cook. 2012. "Creating Open Government Ecosystems: A Research and Development Agenda." *Future Internet* 4 (4): 900–928. https://www.ctg.albany.edu/publications/journals/og_ecosystems_2012/og_ecosystems_2012.pdf.

Heimstädt, Maximilian, Fredric Saunderson, and Tom Heath. 2014. *Conceptualizing Open Data Ecosystems: A Timeline Analysis of Open Data Development in the UK.* Berlin: Freie Universität, School of Business. http://edocs.fu-berlin.de/docs/servlets/MCRFileNodeServlet/FUDOCS_derivate_000000003562/discpaper2014_12-2.pdf.

Hess, Charlotte, and Elinor Ostrom, eds. 2006. *Understanding Knowledge as a Commons: From Theory to Practice*. Cambridge, MA: MIT Press.

Janowski, Tomasz, Jeanne Holm, and Elsa Estevez, eds. 2012. *Open Growth: Stimulating Demand for Open Data in the UK*. Briefing note. Deloitte Analytics. https://www2.deloitte.com/content/dam/Deloitte/uk/Documents/deloitte-analytics/open-growth.pdf.

Janssen, Marijn, Yannis Charalabidis, and Anneke Zuiderwijk. 2012. "Benefits, Adoption Barriers and Myths of Open Data and Open Government." *Information Systems Management* 29 (4): 258–268.

Janssen, Marijn, and Anneke Zuiderwijk. 2014. "Infomediary Business Models for Connecting Open Data Providers and Users." *Social Science Computer Review* 32 (5): 694–711.

Li, Jerry Z., John C. Nesbit, and Griff Richards. 2006. "Evaluating Learning Objects across Boundaries: The Semantics of Localization." *International Journal of Distance Education Technologies* 4 (1): 17–31.

Livingstone, Sonia. 2009. "On the Mediation of Everything: ICA Presidential Address 2008." *Journal of Communication* 59 (1): 1–18.

MacKinnon, Rebecca, Elonnai Hickok, Allon Bar, and Hae-in Lim. 2014. *Fostering Freedom Online: The Role of Internet Intermediaries*. Paris: UNESCO. http://unesdoc.unesco.org/images/0023/002311/231162e.pdf.

Magalhaes, Gustavo, Catarina Roseira, and Sharon Strover. 2013. "Open Government Data Intermediaries: A Terminology Framework." In *ICEGOV 2013, Proceedings of the 7th International Conference on Theory and Practice of Electronic Governance*, 330–333. New York: Association of Computing Machinery.

Meng, Bingchun, and Fei Wu. 2013. "Commons/Commodity: Peer Production Caught in the Web of the Commercial Market." *Information Community & Society* 16 (1): 125–145.

Mutuku, Leonida, and Christine M. Mahihu. 2014. *Open Data in Developing Countries: Understanding the Impacts of Kenya Open Data and Services*. Nairobi: iHub Research. http://www.opendataresearch.org/sites/default/files/publications/ODDC%20Report%20iHub.pdf.

Open Data Handbook. n.d.. Cambridge: Open Knowledge Foundation. http://opendatahandbook.org/guide/en/what-is-open-data/.

Plotkin, David. 2013. *Data Stewardship: An Actionable Guide to Effective Data Management and Data Governance*. Amsterdam: Morgan Kaufmann.

Pollock, Rufus. 2011. "Building the (Open) Data Ecosystem." *Open Knowledge International Blog* (blog), March 31. https://blog.okfn.org/2011/03/31/building-the-open-data-ecosystem/.

Rose, Frank. 1999. *The Economics, Concept, and Design of Information Intermediaries: A Theoretic Approach*. Heidelberg: Physica-Verlag.

Rosenbaum, S. 2010. "Data Governance and Stewardship: Designing Data Stewardship Entities and Advancing Data Access." *Health Services Research* 45 (5), pt. 2: 1442–1455.

Sein, Maung K. 2011. "The 'I' between G and C: E-government Intermediaries in Developing Countries." *Electronic Journal of Information Systems in Developing Countries* 48 (1): 1–14. https://onlinelibrary.wiley.com/doi/epdf/10.1002/j.1681-4835.2011.tb00338.x.

Sein, Maung K., and Bjørn Furuholt. 2012. "Intermediaries: Bridges across the Digital Divide." *Information Technology for Development* 18 (4): 332–344.

Smith, Matthew L. 2014. "Being Open in ICT4D." *Social Science Research Network*, November 17. http://papers.ssrn.com/sol3/papers.cfm?abstract_id=2526515.

Smith, Matthew L., Laurent Elder, and Heloise Emdon. 2011. "Open Development: A New Theory for ICT4D." *Information Technologies and International Development* 7 (1): iii–ix. https://itidjournal.org/index.php/itid/article/viewFile/692/290.

Smith, Matthew L., and Ruhiya Seward. 2017. "Governing Openness in an Unequal World." Ottawa: IDRC, mimeo.

Sorrentino, Maddalana, and Bjoern Niehaves. 2010. "Intermediaries in E-inclusion: A Literature Review." In *Proceedings of the 43rd Hawaii International Conference on System Sciences*, Honolulu, HI, January 5–8. IEEE.

Spence, Randy, and Matthew L. Smith. Forthcoming. *ICTs and Open Provision: Changing Economic and Development Perspectives*. Ottawa: IDRC.

Srnicek, Nick. 2017. *Platform Capitalism*. Cambridge: Polity Press.

Tauberer, Joshua. 2014. "Open Government, Big Data and Mediators." Open Government Data. https://opengovdata.io/2014/open-government-big-data-mediators/.

Tennison, Jeni. 2015. "Why Is Open Data a Public Good?" *Open Data Institute* (blog), February 25. https://oldsite.theodi.org/blog/why-is-open-data-a-public-good.

Tyson, Jeff. 2015. "Global Development's 'Data Intermediaries' and Their Critical Role Post-2015." *Devex*, July 17. https://www.devex.com/news/global-development-s-data-intermediaries-and-their-critical-role-post-2015-86511.

van den Broek, Tijs A., Anne F. van Veenstra, and Erwin J. A. Folmer. 2014. "Walking the Extra Byte: A Lifecycle Model for Linked Open Data." In *Pilot Linked Open Data Nederland, Deel 2—De Verdieping. Linked Open Data*, 98–114. https://research.vu.nl/en/publications/walking-the-extra-byte-a-lifecycle-model-for-linked-open-data.

van Schalkwyk, François, Michael Cañares, Sumandro Chattapadhyay, and Alexandre Andrason. 2015. *Open Data Intermediaries in Developing Countries*. Berlin: Web Foundation. http://webfoundation.org/wp-content/uploads/2015/08/ODDC_2_Open_Data_Intermediaries_15_June_2015_FINAL.pdf.

Wagner, John A. 2013. "Accelerating Biomedical Research through Open Knowledge Creation and Stewardship." *Clinical Pharmacology & Therapeutics* 93 (6): 476–478.

Weber, Steven. 2005. *The Success of Open Source*. Cambridge, MA: Harvard University Press.

3 Trust and Open Development

Anuradha Rao, Priya Parekh, John Traxler, and Rich Ling

Introduction

This chapter focuses on the role of trust in open development initiatives, where open development refers to "the free, public, networked sharing of information and communication resources toward a process of positive social transformation" (Bentley, Chib, and Smith, chapter 1, this volume). In this chapter, we focus on development initiatives that apply open practices in the domains of education (specifically, open education resources) and urban service delivery. Both sectors have key social functionalities that are being increasingly digitalized.

We argue that the success of an open development initiative hinges on trust; that is, the reliance on and confidence in the integrity of the development initiative, the stakeholders, and the digital infrastructure. Trust is a key factor for the various actors who engage with the initiative. Open development, through its various instantiations, has the potential to facilitate access to and use of a wide variety of services and information in the Global South. Nonetheless, it will falter unless users, developers, and other stakeholders have a basic trust in the digital content produced and distributed by the initiative and in the processes of content production itself.

It is important to note that in this chapter we use the term *open system* differently than in systems or in computer science literature. We use it to refer to a digital infrastructure that is used for collecting and/or sharing information from or with a public. We refer to an open system because of the way in which digital infrastructures are utilized within development initiatives, which may contain multiple layers of open processes. For example, a development initiative that collects information from the public may utilize a platform such as Ushahidi, which is itself an open source software package that can be customized through peer production. At the next level, the Ushahidi platform is used within an organizational structure and can be used for collecting information from a public. The sponsoring organization can then perhaps curate the material and share it with the contributors (for example, crowdsourcing, particularly in the noncommercial sense of the word). In addition, the sponsoring organization can use the information and share it with stakeholders (for example, local authorities) or use it in their political project vis-à-vis the stakeholders. Thus, our ideal system is open at several levels (peer production, crowdsourcing, sharing, use in advocacy).

Implicit in a development initiative is the idea that a lack of trust can seriously impede social transformation outcomes of the initiative in question (Diallo and Thuillier 2005; Klijn, Edelenbos, and Steijn 2010). For example, within open development initiatives, users can curtail participation in an initiative if they are unsure whether the digital infrastructure, such as an open source software program, is plagued by computer viruses or is vulnerable to hacking. Users can also doubt the validity of information if they suspect that other stakeholders have corrupted or gamed the material on the system (for example, a nongovernmental organization [NGO] that only presents certain crowdsourced testimonies that speak to their cause). Crowdsourcing may take advantage of collective intelligence, but it also may simply reflect the biased perspective of a faction of users or may reinforce existing privilege and authority (Graham and Haarstad 2013; Singh and Gurumurthy 2013). Any open development initiative involving personal or sensitive information can also falter if users fear for the confidentiality of their information shared on open systems. These and other critical aspects of openness are addressed in the following discussion.

First, we explore concepts of trust and the topic of trust within information and communications technologies for development (ICT4D) and *open* research. We note significant gaps in research related to trust and open development. Next, we contextualize the main trust and open development issues within our chosen areas of open educational resources (OER) and urban services delivery and summarize key similarities across areas. Finally, we conclude by drawing together these insights and proposing a trust model that serves to facilitate the exploration of trust in open development. Our model offers a relational view of trust that can be used to overcome trust issues hindering the contribution of openness toward a process of positive social transformation.

Literature Review Regarding "Trust"

For the purposes of this discussion, trust is viewed as the willingness of an individual to rely on the actions and attitudes of others with regard to future actions. This implies that the person (or group) who trusts gives over some elements of control to the person (or group) being trusted. The person who trusts is in some ways dependent on, and is vulnerable to, the person or group being trusted. They are, in a sense, gambling that the situation will have the resolution they want, but the outcome may indeed result in some harm to the person who trusts. Cognate concepts include notions of reciprocity control, confidence, risk, meaning, and power (Ostrom and Walker 2003).

Trust is an inherently social attribute that permeates various aspects of our social lives. However, trust is intangible and defies easy explanation. A vast and interdisciplinary body of scholarship exists on the meaning, role, and nature of trust in modern societies (Harper 2014a; McKnight and Chervany 2001). Accordingly, the conceptual confusion regarding trust arises from the multiplicity of definitions across disciplines as well as in everyday usage. Trust has been a focus of inquiry in the humanities and social sciences, ranging from consideration by the ancient Greek philosophers (Johnstone 2011; O'Hara 2004) to the contemporary world and the transformative changes that have accompanied the advent of information and communication technologies (ICTs).

Trust does, of course, exist in a wider system of choice, authority, risk, habit, and other factors that determine whether trust is necessary, possible, or even worth the effort. Trust also exists in specific cultural contexts (Zaheer and Zaheer 2006). Thus, any discussion of trust must recognize the impact of culture. Clearly, we exist at the intersection of various cultures—for example, regional, national, religious, and digital—and this intersection will be characterized by specific attitudes. To borrow loosely from Hofstede's (1983) work, this can include dimensions of culture, such as attitudes toward risk taking and risk aversion, short-term and long-term orientation, authority and consensus, individualism and communalism, and so on. The issue of trust implies a certain equality between partners. If the power differential becomes too great, then the interaction is not necessarily trust but rather coercion. That is, the party that is carrying out the task does it based on the threat, not mutual confidence. Also, the operation of trust will be affected by the development initiative, in part because the consequences in various dimensions, such as health care, community education, and local government, might be very different from each other.

Trust and New Information and Communication Technologies

Given the rise of distant interaction, we are faced with the need to evaluate the reliability of the information we receive. A greater portion of the world's population, for example, is faced with the need to interact with remote actors, provide banking information online or through mobiles, gather information on global events, follow election news, and perform a variety of other online actions that rely on our willingness to trust others and the information systems that we use to carry out these activities (Albright 2016; Starbird et al. 2014). Indeed, the recent issue of how much users trust information on social networking sites and how these sites have been used to spread fake news during recent presidential elections in the United States underscores this issue (Tandoc et al. 2018). Thus, it is useful to turn the research spotlight on the notion of trust within ICT-mediated contexts.

Several explanations have been posited for the growing interest in the role and nature of trust in the use of new technologies and online environments. Ess (2010) has pointed out that trust in online environments has been a major focus of attention in part because of the *moral panic* model accompanying new technologies in the modern Western world. He highlighted possibilities for, as well as challenges to, the development and sustenance of trust in online environments. This is particularly the case with the blurring of real and virtual boundaries, which raise new and uncomfortable questions related to trust and virtue ethics. Taddeo (2010) has noted that as we move away from face-to-face interaction and with the growing dependence of users on informational artifacts, earlier questions about trust in technology have been extended into the new digital environments. These include questions about the nature of trust, the requirements necessary for its occurrence, whether trust as an artifact can be developed, or whether trust is reserved for interactions with other humans. Harper's (2014b) book explores the question of trust vis-à-vis computers and the Internet from technical, sociophilosophical, and design perspectives. By combining academic perspectives with concerns raised outside academia, the book raises pertinent questions regarding trust against the backdrop of humanity's ever-increasing *dialogues* with computers.

Trust and ICTs in the Global South

In the specific context of the Global South, Smith (2007) has examined the role of trust in electronically mediated government services in Chile. His work distinguishes between the subjective element of trust (that is, the attitude of the user) and objective characteristics of trustworthiness (that is, of the system relying on the trust). Whereas Smith's emphasis was on building institutional trust through e-government, Morawczynski and Miscione (2008) examined a broader spectrum of trust in the use of the mobile—that is, cell phone—banking system M-PESA in Kenya. As trust was examined in the context of exchange regarding mobile banking, involving a variety of social actors, it comprised several levels: interpersonal trust, such as between the parties exchanging payment; extended trust, which means beyond the individuals who are known personally and also includes others who should have and should not have access to the system; and institutional trust, between individuals and institutions. Srinivasan's (2007) study of an information kiosk project in southern India found personal and institutional trust crucial in determining service usage. However, she noted that existing practices, power relations, and issues of class, caste, and gender that structured the community also shaped perceptions of trustworthiness of the mode and medium of interaction.

Trust in Open Processes

In general, trust has been inadequately dealt with within the literature on open development. The work of Loudon and Rivett (2013) was a first step in this research area, as they highlighted the role of trust relationships in the cocreation of collaborative ICT and development research, as well as the system design and implementation of an e-pharmacy system in South Africa. Literature on open processes has tended to focus on those based on open source software, which has grown into a successful and mainstream movement in and of itself (Feller and Fitzgerald 2000). Although studies have focused on the motivations for (e.g., Davidson et al. 2014; Roberts, Hann, and Slaughter 2006) and the barriers to (e.g., Hars and Ou 2002; Steinmacher, Silva, and Gerosa 2014) participation in open source software projects, the concept of trust is rarely mentioned. The few references to trust include Feller and Fitzgerald's (2000) brief discussion of the role of trust in the marketing of open source software and Olleros's (2008) study of trust in the crowdsourced development of Wikipedia. Two possibilities for the lack of references to trust with regard to open source software projects were highlighted by Gallivan (2001). The first was that trust may be implicit, unacknowledged, and taken for granted in many open source software project activities. The second possible explanation was that most open source software projects relied on a variety of implicit or explicit control techniques and deliberately avoided relying on trust because this could make them vulnerable to members' misdeeds.

Producing and Sharing Open Educational Resources

In the past fifteen years, the OER movement has been gaining traction across the institutions of the developing world. It has inspired greater academic interest in the economic, political, and cultural complexities of this phenomenon. We also note that several other factors can play a key role in fostering trust in sharing OER or in education more generally. These include formal versus informal learning, perceived legitimacy and authority of institutions, and trust in teachers or facilitators and the pedagogies they espouse and enact. Additional factors include the types of technologies and the pedagogies designed or implied in them, levels of digital literacy, and the role and extent of cultural dimensions influencing trust. In this section, we focus specifically on OER because of our focus on open practices, by which we mean how educational resources are produced and shared publicly to benefit a process of positive social transformation.

In the mainstream, OER have reached a clearly defined and documented status such that many institutions may recognize them in a particular way. For example, the United Nations Educational, Scientific and Cultural Organization (UNESCO) defines OER as "teaching, learning and research materials in any medium, digital or otherwise, that reside in the public domain or have been released under an open license that permits no-cost access, use, adaptation, and redistribution by others with no or limited restrictions" (UNESCO 2012, 1). OER include full courses, course materials, modules, textbooks, streaming videos, tests, software, and any other tools, materials, or techniques used to support access to knowledge.[1]

Since the definition of these educational resources in 2002, there has been an upsurge in interest in their development and use. Indeed, higher education institutions (HEIs), intergovernmental organizations, and NGOs located predominantly in the Global North have sought to expand access to and improve the quality of education in developing countries through the introduction of OER. These open materials are seen as a mechanism to address some of the formidable educational challenges in the Global South, which include unequal access to education; variable quality of educational resources, teaching, and student performance; and increasing cost and concern about the sustainability of education (Arinto et al. 2017). Yet, since much of the literature either is devoted to the Global North or fails to adequately consider the local context into which OER are adapted, evidence of the impact of OER in developing countries is still lacking (Arinto et al. 2017).

Trust has figured only marginally in these discussions, although mainly in relation to open education in the developed world. Preliminary research on trust in open education resources has focused on the reliability and quality of materials (Peacock, Fellows, and Eustace 2007); trust in the validity of open learner system evaluations (Ahmad and Bull 2008); trust mechanisms in the search and reuse of OER by teachers (Clements and Pawlowski 2012); privacy and security concerns; and the sharing attitudes of learners in online personal learning environments (Tomberg 2013).

However, we must also recognize that people and communities, particularly those in developing countries, continue to undertake learning that they themselves value, often using digital resources, in ways that do not fit within the strict definition of an OER used within the mainstream. This is why a context-driven and practice-based lens is needed to explore issues of trust in the sharing and production of OER. We illustrate how trust may be affected differently based on three different contexts within which OER have been shared and produced to work toward a process of positive social transformation. The first is found within traditional higher education institutions, such as the production and sharing of OER between teachers. The second occurs outside a traditional learning environment, such as within an NGO's initiative to build OER to solve a specific development problem. The third takes the learners' perspective and considers sharing and producing OER through both formal and informal channels. In each of these cases, the role of trust factors into the discussion differently because of the stakeholders involved and the ultimate goals of the open system.

Within a traditional HEI context, OER have the *potential* to impact teaching and learning in several ways, including improvement in student performance and satisfaction, more equitable access to education, critical reflection by educators, and financial benefits for students and/or institutions (Weller et al. 2015). Recent research has acknowledged that OER can also positively influence educators' pedagogical perspectives and practices, and the resultant empowerment of students and educators could disrupt the power dynamics traditionally associated with the transmissive educator/student relationship (Arinto, Hodgkinson-Williams, and Trotter 2017; Hodgkinson-Williams et al. 2017). However, these studies also note that such phenomena are in a nascent stage in the Global South, given the cultural and infrastructural constraints within which conventional learning occurs. These constraining factors affect how OER are perceived and incorporated by institutions, teachers, and learner communities—an aspect that is discussed in more detail later in this section.

Furthermore, within a traditional HEI context, there are also different culturally specific conceptions of what constitutes education, even within a specific society. These definitions have different consequences for where and how trust could operate within the HEI context. As an example, to pursue the liberal Western conceptions of education, there are competing alternatives, ranging from learning that grows out of the transmission and absorption of content—that is, information and procedures—to learning that grows out of discussion, which relates to experience or understanding (Conole et al. 2004). Comparing these, the focus of trust is on the source of the content in the former and on the quality of the relationships in the latter (Curzon-Hobson 2002; Sidorkin 2000). Of course, the pedagogy, which is what we are describing here, is not independent of the subject, so where trust needs to be placed depends on the nature and content of what is being learned; in other words, whether we are learning astrophysics or recent political history, whether we are learning apparently established, stable, objective, and universal truths or emergent, partial, local, or contested perspectives.

Whether either of these types of ideas, and OER, comes from within one's own culture or from outside—and we must recognize here the intellectual hegemony and productive force of the Global North—also impacts whether what might be learned will in fact be trusted. For instance, Hodgkinson-Williams et al. (2017) found, in a meta-analysis of thirteen OER projects that took place in developing countries, that educators preferred to use resources *as is*. This suggests that materials emanating from the North can easily enter into curricula from outside one's culture. Yet educators also had difficulty searching through resources, and preferred resources, that were relevant to their context. Trusted sources may be established purely because educators have learned to find content easily through them. This may represent an abiding challenge for OER, that of locating them in repositories of a realistic size from outside the cultural context of their origin. Without the appropriate safeguards in place, there is also the danger that OER can constitute a kind of knowledge or information imperialism (Mulder 2008).

Outside the traditional higher education context, take for example the experience of the AgShare Today Project in Ethiopia and in Uganda. When the production of OER was grounded in practical experience, as well as Indigenous and social knowledge, education and training became more responsive to the agricultural problems facing these countries (Hassen 2013; Kaneene et al. 2013). Multimedia OER, such as case studies, videos, and modules developed through participatory action research (PAR), where academics, small-holder farmers, and NGOs collaborate. The multimedia OER are then shared and used in two ways. First, they can be integrated into a university's curriculum, such that students engage with grounded, authentic materials. Second, they can also be embedded within the NGO's or farming union's interventions and practices, such that the farmers and practitioners can regularly use, modify, and update these materials as well.

However, involving a greater range of actors, such as the NGOs and farmers, in addition to students, teachers, and administrators, introduces complexity where trust is concerned. Usually, for PAR to work well in the first place, strong, trusting relationships must exist (Kemmis and McTaggart 2000). Moreover, the involvement of a sponsoring organization, such as an NGO, as well as whoever agrees to administer and maintain the system on which the OER are stored and distributed, also imply interplays of trust. Not only must the NGO be considered a trustworthy organization for the farmers and educators to participate in the initiative, there must also be clear and consistent oversight over the materials shared through the system in order for educational institutions to continue using them in their courses. Sharing the OER publicly also implicates trust in the content, especially if users wish to take the content and apply it directly within a context completely different from that in which the OER was developed. If the resources are modifiable, they may have also been altered unsuspectingly. To understand how resources are being used and changed, the sponsoring organization may need to monitor activity within the system.

Furthermore, the growth and potential of OER in the Global South must be viewed in relation to the numerous challenges, including the uneven institutional, political, economic, and technological terrain, within which these initiatives are taking place (Kanwar, Kodhandaraman, and Umar 2010; Peters and Britez 2008; Smith 2013). Trust in the quality and usability of these resources, institutions, channels of learning, teachers, and learner communities can play a crucial role in determining the success of OER initiatives in these contexts. They can also be affected by the channel (informal or formal) through which they are diffused (Rogers 2003). In contrast, when OER are used to support other educational goals, such as when a student's sense of agency or social inclusion is targeted through constructivist pedagogy (Hodgkinson-Williams and Paskevicius 2012), the quality of the materials becomes less important than the process through which they were created. In this case, it is far more important for the students to trust their educators and the administrators of the open system in order to feel comfortable contributing.

In sum, we have argued that the trust relations that are important to consider within OER initiatives largely depend on the context within which they occur. There can be trust issues between the users and creators of educational content, but there is also trust hinging on the trustworthiness of the sponsoring organization (the educational institution or NGO), the operators of the open system (how OER are collected and distributed), and the broader institutional environments within which OER initiatives are embedded.

Crowdsourcing and Peer Production for Improving Urban Services

Open systems in the delivery of urban services include projects that, for example, enable public participation, crowdsourcing, and information sharing in order to report and resolve problems with public service delivery in cities. Looking at urban service delivery, deficiencies in public services in urban and periurban areas, particularly in the Global South, have precipitated the rise of ICT-based solutions for improving public services such as sanitation, solid waste management, water supply, and transportation (Bhatnagar 2014; Fang 2014; National Institute of Urban Affairs 2015). These have been joined by a growing number of open development projects that have leveraged crowdsourcing and the potential of photographic evidence, as well as the geolocative functionality of ICTs. Examples include systems to report and resolve issues of transportation or mobility, such as the open data/open source transport systems in the Philippines (World Bank 2013a), Mexico City (World Bank 2013b), and Nairobi (Williams et al. 2015); crowdsourcing systems to measure and share air quality, such as #Breathe, The IndiaSpend Air Quality Index Network (IndiaSpend 2015); and platforms for disaster management and other emergency responses, such as Ushahidi, OpenStreetMap, and Sahana EDEN (de Silva and Prustalis 2010; Ortmann et al. 2011). Other crowdsourcing systems, such as Praja.in in India (Rao 2014; Rao and Dutta 2016) and the Taarifa platform in Zimbabwe (Iliffe et al. 2014), allow citizens to engage with their local governments in the monitoring and reporting of service delivery and other local civic issues.

The main discourse surrounding these crowdsourcing systems has been that crowdsourcing can be used to inform policymakers and thereby work to improve public service efficiency, particularly in the areas of urban mobility (Williams et al. 2015; World Bank 2013a; World Bank 2013b) and disaster management (Camponovoa and Freundschuh 2014; de Silva and Prustalis 2010). An example of this type of system is Praja.in, a blog-based platform in Bangalore, India, intended to develop transparency and accountability in governance.[2] The system includes a website developed and curated by a not-for-profit organization, where individuals can report various types of civic issues (Rao 2014; Rao and Dutta 2016). These issues can range from the disposition of abandoned vehicles, to the planting of trees, to the checking of electrical meters. Here, we note that there are three types of stakeholders, with different responsibilities. First, there are the citizens who provide information to the system, presumably reporting issues of interest to them. This is the input into the system that is administered by the organization. Second, there is an NGO responsible for developing and managing the system, which we call a sponsoring organization. Third, there is an external public that is the target audience for this information. The output includes the crowdsourced development of reports on various issues, which are then submitted to governmental authorities in support of different policies. It can also include information provided back to the publics that use the system itself. In a broad sense, there are different actors implicated in the system that need to act and respond in specific ways for the system to be effective. If trust between these stakeholders is lacking, the system will not be effective.

With regard to open urban service delivery projects, we posit that trust is a critical factor in both the creation and acceptance of open systems. The most significant critiques of government open data initiatives have stemmed from urban studies academics and practitioners in India, as noted by Bhuvaneswari Raman (2012) and Nithya V. Raman (2012). For example, a civil society organization called Transparent Chennai sought to improve urban services for poor and marginalized people in the city of Chennai, as observed by Nithya V. Raman (2012). In India, following the 2005 Right to Information Act (RTI Act), citizens were promised public access to government information in a timely manner. Yet, when Transparent Chennai began collecting information about public toilets, pedestrian walkways, and bus routes to visualize this information more effectively, the government either did not have the information or stalled on their requests. The practitioners had to call, write, and visit government offices on numerous occasions to gain access to the information. When the information was received, there were often parts missing or intentionally omitted as a result of power conflicts between different community groups. Invariably, what Nithya V. Raman (2012) found was that information related to urban services for the poorest and most marginalized groups in Chennai was often disregarded by public officials.

In this example, the most significant trust issue takes place between the information providers, or the government officials, and the users of the information and civil society organizations. However, if the organization then chooses to add to or modify the information provided in order to fill in the gaps, or to right the wrongs so to speak, then the organization subjects itself to significantly different trust issues. If the organization creates open data to share the modified government information, then users of the information, such as citizens and policymakers, must also trust the organization. If the organization uses crowdsourcing to collect information directly from citizens in lieu of modifying government information, it might face a challenge in encouraging government officials to use this information in its policymaking. In other words, there are many dimensions of trust that are important to think about when one considers both the practices of the system (sharing or crowdsourcing public information) and the social transformation outcomes sought (improved delivery of urban services). In the next section, we outline a systematic way to dig deeper into the trust issues in play.

Dimensions of Trust in Open Development: Preliminary Research Directions

Given this background, we are interested in developing a model and outline the research questions that examine the roles of various actors vis-à-vis trust and open systems. The model endeavors to consider the role of trust from different positions. Integrating our analysis of open development initiatives across open education resources and urban services, there are three main types of stakeholders. There are, in general, those who sponsor the system. They may be the same as those who develop the system, though that is not necessary. This group may also curate the information for a particular cause, as in the Praja.in example offered in the previous section. In addition, there are those who contribute their content, observations, complaints, or insights. Finally, there are those who receive the output of the system, such as actors who are charged with policymaking or students who are taught by teachers using OER. This echoes the open production, open distribution, and open consumption categories outlined by Smith and Seward (2017).

Our model outlines trust among these stakeholders in an open development initiative. It includes an overview of the processes through which they cultivate trust and develop trustworthiness cues (see figure 3.1). The model highlights various elements and dimensions of trust, including technical, institutional, and social aspects. Furthermore, an examination of the relationships within and between these elements helps to illuminate the various levels at which trust can develop and reside in open systems.

In ascertaining trust, it is important to identify the types of stakeholders, their roles, and their relationships, along with the interactions that

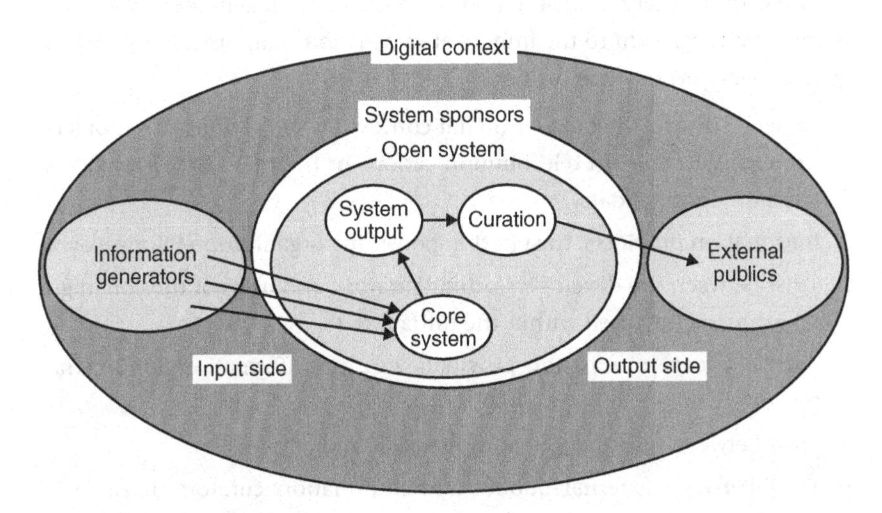

Figure 3.1
A model of trust between stakeholders using open systems and the ecosystem of trust.

foster trust among them. Gallivan (2001) has pointed to three types of trust within a user community, which also reflect the types of users of an open system: trust in the human leadership; trust in the broader community of users who test the software code, identify possible problems, and submit software bug reports; and the trust that developers assume of end users (for example, trust that they will not commercialize the code that they developed).

In the case of OER or urban services in general, key actors vis-à-vis trust would include:

1. Those who generate the information (for example, citizens who document urban infrastructure problems and report them to an advocacy group via a system, or teachers who create OER).
2. Those who develop or administer the core digital infrastructure of the system (which can include the source code of the software or other proprietary or open source tools that facilitate the exchange of information and services, such as email lists, websites, or shared folders).
3. Those who sponsor the system, which includes those who process contributed information, curate it, and present it to external publics.
4. The external publics to whom the information is presented, which includes anyone who may wish to use the information.

There are a variety of trust relations among these stakeholders. Some are more directly relevant to the immediate discussion than others; however, a general overview of these includes:

1. General trust in the broader digital context by the various stakeholders (for example, trust in telecommunications or Internet service providers to guard personal data).
2. Information providers' trust in the sponsoring organization, if one exists.
3. Trust between the developers, administrators, and users of the core digital infrastructure used within the initiative.
4. Trust between information providers and the information administrators or curators.
5. Trust between information administrators and curators.
6. Trust between external publics and information curators, if there is curation.

It is proposed that each type of user will have different issues and different criteria as the individual develops a sense of trust. We outline relevant questions that need to be addressed in examining these issues.

General Trust in Digital Contexts

Before looking at the more specific domains of the stakeholders, it is worth noting that, more broadly, those who contribute information to open systems (as well as those who use it) need to trust what we call the *digital context*. That is, they need to trust the legitimacy of actors such as service providers, governments, and other content providers. Users need to be able to make the assumption that, for example, there is no risk that their data and identity will be hacked, profiled, or misused by unknown third parties. There is the necessity to trust (or tolerate risk) in institutions, such as telecommunications operators or Internet service providers, and their treatment of personal information (Middleton 2012). Users also need to trust that their material will not be stolen or surveilled by hackers or governments (Cox 2014). If an individual on the input side, such as a person who wants to report a street that needs repair, does not have basic trust in the digital context or the technical system, then that individual will not likely use a system to report the problem. If, for example, individuals fear that authorities will be able to trace and persecute them for reporting problems, they may choose not to participate. In other cases, where personal information is not shared and there are no restrictions on how contributed information is used or abused, trust in the digital context may not play a significant role. This is likely the case for users who are just accessing shared information instead of contributing content.

Thus, the broadest research question is, "What dimensions are associated with users' trust in the digital context?"

Information Providers' Trust of System Sponsors

The next issue may be the general trust in the sponsoring institution. Both the people who enter information into the system and those who are asked to base policy on the recommendations of the system need to trust the organization that sponsors the open system (for example, the people providing data to Praja.in and those who are asked to base policy decisions on the material need to have a broader trust in the organization itself, not just in the technical system). These stakeholders need to trust the broader goals, mission, and reputation of the sponsoring groups with whom they share information. Smith (2007) has noted that in the case of governmental institutions, users are influenced by impersonal institutional mechanisms, the people enacting roles and following rules and norms, or the *ethos* of an institution. Levels of trust in an institution are also influenced by its history, satisfaction with its quality and services, and faith in the institution and its leadership (Morawczynski and Miscione 2008). The same is likely to apply to the NGOs or educational institutions that sponsor open systems. For instance, users may prefer to trust massive open online courses (MOOCs) sponsored by universities they deem to have good reputations.

To understand people's notions of trust at this level, it is necessary to answer the questions "What are the ways by which users gauge the reputation or legitimacy of people or leaders and institutions associated with an open system?" and "How does this translate into trusting the particular open system?"

An important aspect of the perceived general trustworthiness of a system is the presence or absence of scandals and fraud on the part of the sponsoring organization in the minds of eventual users. This is true for those who provide information to the system and those who are asked to develop policy based on outputs from the system. For example, ethical questions or scandals associated with open access journals and crowdsourcing sites such as Wikipedia,[3] TripAdvisor, and Yelp (Ferguson, Marcus, and Oransky 2014; Lukić et al. 2014; Tuttle 2014) can color information providers' willingness to trust the sponsoring organization. They can also affect the willingness of external publics to give the open system credence. If, for example, the organization has a reputation for trafficking in partial truths, gaming information, or purveying inappropriate results, it will corrode trust with both information providers and external publics.

Bringing this to the level of open systems in the Global South, to what extent do scams and scandals resulting from malpractice of the sponsoring organization impact the level of trust in an open system?

Trust between Developers, Administrators, and Users of the Digital Infrastructure

The next element of trust addresses the issue of trust between developers and administrators and users in the management and use of the digital infrastructure of the open system. This dimension of trust is not directly associated with information providers' willingness to use the system, but, when considering the broader dimensions of trust, it is part of the larger framing. Open systems typically rely on a digital infrastructure to share resources. Digital infrastructures can range from web-based applications, such as a wiki platform, to email listservs or file-sharing services such as DropBox. In some cases, open systems may grant privileges to users to view and modify the source code of web-based applications or to change or delete folders within a file-sharing system. In this case, one can ask to what extent developers trust *lay* users. To what degree are others given access to, for example, the underlying digital infrastructure? For example, are lay users allowed to access the core code of Ushahidi or Wikipedia? Is it acceptable to take it and develop commercial versions of the software? Alternatively, do those who develop the system have some type of veto power (Olleros 2008; Sfakianakis et al. 2007)? Is there a pathway for a user to become a core developer or administrator? In other words, are lay users trusted enough by the core group of developers to share and/or modify the digital infrastructure? Taking this further, can there be situations in which there is user-developer interaction in system design and implementation? How welcome is this input? It is also possible to ask through what processes and interactions mutual trust is built and exercised in these situations (Loudon and Rivett 2013).

Trust between Users Who Input Information and the Open System

The next aspect of trust is related to the decision by information providers to contribute content to the system (which we term *input trust*). For example, what are the trust-based antecedents of a person who reports a pothole in the street or, for that matter, a corrupt official, on a site such as Praja.in? When examining the situation from the perspective of the information providers, the question is whether they trust that the system will record their input without resulting in unforeseen negative repercussions, some of which are discussed in the following section.

This moves beyond the more macro issues of trust noted earlier and focuses on the micro aspects that users face in their day-to-day interactions with the system. Considerations can include issues such as whether an individual will encounter computer viruses, increase their chance of being hacked, or have their private information exposed without permission while using the system. When thinking of closed commercial sites, there are examples, such as that of the AshleyMadison.com scandal, where the identities of people using this dating site—designed for people who are otherwise in committed relationships—were hacked. One can imagine that similar exposure could take place on sites where citizens report corrupt officials or, for example, with HarassMap, a site that compiles reports of violence in a given neighborhood. If the eventual victim of harassment fears that she or he will be unmasked online, meaning that they do not trust the reporting system, they will not use it.

This echoes William and Dorothy Thomas's (1928) notion that if one believes situations are real, they are real in their consequences. Regardless of whether open systems are more vulnerable to this thinking, it is possible to postulate that users will have this fear. This issue is perhaps more relevant in the case of open systems, since they are often developed outside formal frameworks, such as nondisclosure agreements, common legal entities that can be held liable, or other traditional institutional arrangements. This is more specific than the broader issue of trust in the wider digital context noted earlier. Thus, a relevant question is, to what extent do users who provide information for an open system trust the security of the specific system?

Conversely, another possible factor affecting how input trust develops could stem from the perceived motivations of co-users of the system. Studies have highlighted that altruism, internal values, and community identification are some key motivations for participants who contribute to open source software projects (Davidson et al. 2014; Roberts, Hann, and Slaughter 2006). Thus, we ask, "Do (and, if so, how and for whom) perceptions of the complementarity or mutual reinforcement of motivations facilitate the development of trust among contributors to open systems?"

There appears to be a reciprocal dimension to trust in crowdsourced systems, where the information providers are both the source of individual reports and the beneficiaries of aggregated reports or analysis. This is seen in systems such as driving and traffic apps that show traffic jams. In this case, as individuals provide their personal situations and receive systemic information, there is the need for users to ascertain the trustworthiness of the information.

Another dimension is to look at trust from the perspective of the system sponsors, who need to be sure that people who input items are providing legitimate information. One threat to the validity of the information can be when the institutions that are evaluated by the system *shape* the data providing false inputs. For example, the highway authority may wish to water down reporting of potholes by posing as users and giving their organization high positive evaluations. In effect, they can act as *gamers* of the system, who reduce the reliability of the information and ratings. In other cases, there can be a perceived need to moderate the incoming material to avoid ranting by users. Thus, we ask, "Are there mechanisms in place to ensure the accuracy of input information, such as systems of moderation, editing, or posting rules?"[4] and "When do users implicitly trust input information as being accurate, reliable, and useful?"

Trust in System Administrators and Curators

Credibility of information in an open system is also related to factors such as the presence of social validation or feedback and the development of collective judgments (Jessen and Jørgensen 2012). Research indicates that trust is enhanced in open system communities when users are involved in the quality assurance process via commenting on, recommending, tagging, and rating resources (Clements and Pawlowski 2012). Thus, we ask, "When and for whom do peer reviews, ratings, and recommendations influence trust in information online?"

Credibility of online information is also dependent on *trustees*, who may or may not be experts but act as a form of authority, having a known identity or profile (Jessen and Jørgensen 2012; Resnick et al. 2000). Therefore, reputation, which is based on a continued identity, as well as a history of performances associated with that identity (Turilli, Vaccaro, and Taddeo 2010), can influence perception, use, and reuse of informational resources (Clements and Pawlowski 2012). For example, the project PetaBencana.id, which reports on flooding in Jakarta, was developed by NGOs and has become a part of the Indonesian National Disaster Management Agency's tools to deal with this issue. Therefore, we ask, "For whom, and to what extent, does the reputation of organizations, institutions, or online identities influence trust in online information and resources?"

At another level, there is the reputation of the informational artifact or the system itself. In this instance, users might *key in* their trust in the system based on trustworthiness cues (Smith 2007). Design and the role of designers as they create new features and user interfaces can play a part in fostering and sustaining trustworthy online services (Cheshire et al. 2010). As most ordinary users cannot make complex technical decisions about the selection of trustworthy software components, trust vis-à-vis the designers selected by administrators of the system is critical (Cheshire et al. 2010; Clark 2014). In order to examine this, we ask, "To what extent, and for whom, do different design cues influence user trust, and how do designers foster and sustain trust among different users of a system?" and "What, and for whom, are the other salient trustworthiness cues exhibited by the system, and in what ways do they influence user trust?"

Trust by External Publics

At the level of output information, the issue of trust arises in relation to whether collected or curated information from open systems is both reliable (and at least testable) and actionable. Reliability is threatened when, for example, there is the sense that the information has been cherry-picked by the sponsoring agency to produce a particular outcome or when there is the sense that stakeholders have been "gaming the system" (Cherry 2013; Notenboom 2014). Such perceptions will reduce the trustworthiness of the information in the eyes of the public. System sponsors will need to adopt various content moderation strategies to deal with these types of fraud (Luca and Zervas 2016). Thus, we ask, "What procedures are in place to ensure that output information is not gamed in an open system?" and "How do users navigate the complexities of gaming the system, and how does this influence their perceptions of a system's trustworthiness?"

Next, output information is deemed more actionable if it is seen as trustworthy; that is, it can be used or easily modified within learning and teaching practices without having to check facts or redesign the resource completely. In the case of crowdsourced urban services, the situation can be somewhat more complex, as various types of advocates, who are pushing a position or pursuing a certain agenda, also gather information. The existence of parallel systems could be resented by those who have status or an investment in the open system and do not wish to change the status quo. In these situations, the legitimacy of the information—and, by extension, trust in the system—will often become a central element in how actionable the information might be. If there is a system of grafting and corruption associated with a particular public service, there will also be a strong motivation to discount the information from the open system since it is a threat to the status quo. In this instance, we argue that the existing relationships and types of stakeholders within the external public category are important to differentiate. We ask, "How do government, NGOs, corporate, and higher education institutions that use or are presented with information determine its trustworthiness?" and "How, and to what extent, does the nature of the relationship between these institutions and the open system (confrontational, cooperative, or a combination) influence their perceptions of trustworthiness of output information?"

The dimensions of trust identified here relate to different components of the open system proposed by the authors and can also be discussed in terms of a trajectory or "career of trust" (Sztompka 1999, 14). In other words, the model can help us identify how trust in a particular open system is built up and sustained over a period of time, the factors that encourage and impede trust, how trust is lost, and what can be done to regain trust in an open system. This last aspect is significant because, as the possibility of gaming the system increases in the context of more technical and automated systems (Ferguson, Marcus, and Oransky 2014), it is important to consider what measures are taken to prevent the recurrence of loss of trust in a system. As a result, we ask, "What are the processes of trust repair and in what ways can trust and reputation be recovered?" and "What, if any, situations exist under which lost trust can or cannot be regained?"

In summary, we have developed a model that illuminates the different stakeholders concerned with the development and use of open systems. Based on this, we have examined the interactions between the different actors vis-à-vis the role of trust and the role that trust plays in the functioning of the system.

Conclusion

This chapter has developed a model and laid the groundwork for the examination of trust expressed by users and developers of open systems used in initiatives in education and urban services delivery. Through an examination of the processes and practices adopted by sponsors of open systems to cultivate trust and develop trustworthiness cues, we posit that the framework will help in understanding the efficacy of such systems. We also highlight some critical perspectives of openness, raising questions with regard to the value, desirability, and altruistic associations of the concept itself. Singh and Gurumurthy (2013) have noted potential problems arising from the uncritical system of the concept of openness to development and argued instead for the establishment of *public-ness*. Song (2015) has highlighted the danger that openness could become a type of orthodoxy that is difficult to question. To be successful, he stressed, a development initiative must provide the *choice* to question it, which, in turn, nurtures reciprocity and trust.

In the domain of education, the privatization and commercialization of universities produces tensions for the open education movement. Trust, authority, respect, loyalty, esteem, and legitimacy are consequently also problematic. Open urban service initiatives in developing countries could be critiqued for their inherent assumptions of digital access and empowerment. Further, if data generated and coproduced on an open system is skewed toward a certain socioeconomic demographic, this could raise doubts about the trustworthiness of its informational resources. By highlighting these issues, this chapter paves the way for a more critical examination of the role of trust and openness in open education and urban services delivery projects. Furthermore, through an exploration of the micro and macro aspects of trust that are highlighted in the chapter, it is hoped that researchers will engage in more nuanced empirical examinations of trust in open development initiatives.

In the cases of open education and urban services, it is worth using the considerations developed here to understand whose interests are being served through the development and use of open systems. Are these systems effective at channeling the needs of citizens and students into the policymaking arena? Alternatively, are there other social forces, such as trust issues, that discount the efficacy of the systems? The model developed here and the research issues that are outlined can help to address these problems.

Notes

1. For additional information, see https://www.hewlett.org/strategy/open-educational-resources/.

2. See http://praja.in/.

3. Such as the illicit edits to the pages of the former British prime minister Gordon Brown regarding drug abuse and criminal activities.

4. See, for example, https://www.ihub.co.ke/ihubresearch/jb_VsReportpdf2013-8-29-07-38-56.pdf.

References

Ahmad, Norasnita, and Susan Bull. 2008. "Do Students Trust Their Open Learner Models?" In *Adaptive Hypermedia and Adaptive Web-Based Systems*, edited by Wolfgang Nejdl, Judy Kay, Pearl Pu, and Eelco Herder, 255–258. Berlin: Springer-Verlag.

Albright, Jonathan. 2016. "The #Election2016 Micro-Propaganda Machine." *Medium*, November 18. https://medium.com/@d1gi/the-election2016-micro-propaganda-machine-383449cc1fba#.corbvpm9s.

Arinto, Patricia B., Cheryl Hodgkinson-Williams, Thomas King, Tess Cartmill, and Michelle Willmers. 2017. "Research on Open Educational Resources for Development in the Global South: Project landscape." In *Adoption and Impact of OER in the Global South*, edited by Cheryl Hodgkinson-Williams and Patricia B. Arinto, 3–26. Cape Town: African Minds; Ottawa: IDRC. https://doi.org/10.5281/zenodo.1005330.

Arinto, Patricia B., Cheryl Hodgkinson-Williams, and Henry Trotter. 2017. "OER and OEP in the Global South: Implications and Recommendations for Social Inclusion." In *Adoption and Impact of OER in the Global South*, edited by Cheryl Hodgkinson-Williams and Patricia B. Arinto, 577–592. Cape Town: African Minds; Ottawa: IDRC. https://doi.org/10.5281/zenodo.1043829.

Bhatnagar, Subhash. 2014. "Public Service Delivery: Role of Information and Communication Technology in Improving Governance and Development Impact." ADB Economics Working Paper Series 391. Manila: Asian Development Bank. https://www.adb.org/sites/default/files/publication/31238/ewp-391.pdf.

Camponovoa, Michael E., and Scott M. Freundschuh. 2014. "Assessing Uncertainty in VGI for Emergency Response." *Cartography and Geographic Information Science* 41 (5): 440–455.

Cherry, Steven. 2013. "Can You Trust an Amazon Review? Reviewers Are Gaming the System at Amazon and Elsewhere for Mischief, Politics, and Profit." *IEEE Spectrum*, February 19. https://spectrum.ieee.org/podcast/geek-life/tools-toys/can-you-trust-an-amazon-review.

Cheshire, Coye, Judd Antin, Karen S. Cook, and Elizabeth Churchill. 2010. "General and Familiar Trust in Websites." *Knowledge, Technology & Policy* 23 (3–4): 311–331. https://link.springer.com/article/10.1007/s12130-010-9116-6.

Clark, David. 2014. "The Role of Trust in Cyberspace." In *Trust, Computing, and Society*, edited by Richard H. R. Harper, 17–37. Cambridge: Cambridge University Press.

Clements, Katie I., and Jan M. Pawlowski. 2012. "User-Oriented Quality for OER: Understanding Teachers' Views on Re-use, Quality, and Trust." *Journal of Computer Assisted Learning* 28 (1): 4–14.

Conole, Gráinne, Martin Dyke, Martin Oliver, and Jane Seale. 2004. "Mapping Pedagogy and Tools for Effective Learning Design." *Computers & Education* 43 (1–2): 17–33.

Cox, Joseph. 2014. "Why Telecom Companies Are Getting Transparent on Government Surveillance." *Motherboard,* June 11. https://www.vice.com/en_us/article/nze79b/why-telecoms-companies-are-getting-transparent-on-government-surveillance.

Curzon-Hobson, Aidan. 2002. "A Pedagogy of Trust in Higher Learning." *Teaching in Higher Education* 7 (3): 265–276.

Davidson, Jennifer L., Umme A. Mannan, Rathika Naik, Ishneet Dua, and Carlos Jensen. 2014. "Older Adults and Free/Open Source Software: A Diary Study of First-Time Contributors." In *OpenSym '14: Proceedings of the International Symposium on Open Collaboration,* 1–10. New York: Association for Computing Machinery.

de Silva, Chamindra, and Mark Prustalis. 2010. "The Sahana Free and Open Source Disaster Management System in Haiti." In *ICT for Disaster Risk Reduction: ICTD Case Study 2,* edited by UN-APCICT/ESCAP, 110–124. Incheon: Asian and Pacific Training Centre for Information and Communication Technology for Development and United Nations Economic and Social Commission for Asia and the Pacific.

Diallo, Amadou, and Denis Thuillier. 2005. "The Success of International Development Projects, Trust and Communication: An African Perspective." *International Journal of Project Management* 23 (3): 237–252.

Ess, Charles M. 2010. "Trust and New Communication Technologies: Vicious Circles, Virtuous Circles, Possible Futures." *Knowledge, Technology & Policy* 23 (3–4): 287–305.

Fang, Ke. 2014. "'Smart Mobility' for Developing Cities." *World Bank Transport for Development* (blog), December 31. http://blogs.worldbank.org/transport/smart-mobility-developing-cities.

Feller, Joseph, and Brian Fitzgerald. 2000. "A Framework Analysis of the Open Source Software Development Paradigm." In *ICIS '00: Proceedings of the Twenty First International Conference on Information Systems,* Brisbane, Australia, December 10–13, 2000, edited by Wanda J. Orlikowski, 58–69. Atlanta: International Conference on Information Systems.

Ferguson, Cat, Adam Marcus, and Ivan Oransky. 2014. "Publishing: The Peer-Review Scam." *Nature* 515 (7528): 480–482. http://www.nature.com/news/publishing-the-peer-review-scam-1.16400.

Gallivan, Michael J. 2001. "Striking a Balance between Trust and Control in a Virtual Organization: A Content Analysis of Open Source Software Case Studies." *Information Systems Journal* 11 (4): 277–304.

Graham, Mark, and Håvard Haarstad. 2013. "Open Development through Open Consumption: The Internet of Things, User-Generated Content and Economic Transparency." In *Open Development: Networked Innovations in International Development,* edited by Matthew L. Smith and Katherine M. A. Reilly, 79–111. Cambridge, MA: MIT Press; Ottawa: IDRC. https://idl-bnc-idrc.dspacedirect.org/bitstream/handle/10625/52348/IDL-52348.pdf?sequence=1&isAllowed=y.

Harper, Richard H. R. 2014a. "Introduction and Overview." In *Trust, Computing, and Society*, edited by Richard H. R. Harper, 3–14. Cambridge: Cambridge University Press.

Harper, Richard H. R. 2014b. *Trust, Computing, and Society*. Cambridge: Cambridge University Press.

Hars, Alexander, and Shaosong Ou. 2002. "Working for Free? Motivations for Participating in Open-Source Projects." *International Journal of Electronic Commerce* 6 (3): 25–39.

Hassen, Jemal Y. 2013. "The Potential of a Multimedia Open Educational Resource Module in Enhancing Effective Teaching and Learning in a Postgraduate Agricultural Program: Experience from AgShare Project Model." *Journal of Asynchronous Learning Networks* 17 (2): 51–62. https://files.eric.ed.gov/fulltext/EJ1018270.pdf.

Hodgkinson-Williams, Cheryl, Patricia B. Arinto, Tess Cartmill, and Thomas King. 2017. "Factors Influencing Open Educational Practices and OER in the Global South: Meta-Synthesis of the ROER4D Project." In *Adoption and Impact of OER in the Global South*, edited by Cheryl Hodgkinson-Williams and Patricia B. Arinto, 27–67. Cape Town: African Minds; Ottawa: IDRC. https://doi.org/10.5281/zenodo.1037088.

Hodgkinson-Williams, Cheryl, and Michael Paskevicius. 2012. "The Role of Postgraduate Students in Co-authoring Open Educational Resources to Promote Social Inclusion: A Case Study at the University of Cape Town." *Distance Education* 33 (2): 253–269.

Hofstede, Geert. 1983. "National Cultures in Four Dimensions: A Research-Based Theory of Cultural Differences among Nations." *International Studies of Management & Organization* 13 (1–2): 46–74.

Iliffe, Mark, Giuseppe Sollazzo, Jeremy Morley, and Robert Houghton. 2014. "Taarifa: Improving Public Service Provision in the Developing World through a Crowd-sourced Location Based Reporting Application." *Journal of the Open Source Geospatial Foundation* 13 (1): 34–40. https://journal.osgeo.org/index.php/journal/article/view/211.

IndiaSpend. 2015. "Introducing #Breathe, The IndiaSpend Air Quality Index Network." *IndiaSpend*, December 1. http://www.indiaspend.com/cover-story/introducing-breathe-the-indiaspend-air-quality-index-network-39580.

Jessen, Johan, and Anker Helms Jørgensen. 2012. "Aggregated Trustworthiness: Redefining Online Credibility through Social Validation." *First Monday* 17 (1). https://firstmonday.org/article/view/3731/3132.

Johnstone, Steven. 2011. *A History of Trust in Ancient Greece*. Chicago: University of Chicago Press.

Kaneene, John B., Paul Ssajjakambwe, Stevens Kisaka, RoseAnn Miller, and John D. Kabasa. 2013. "Creating Open Education Resources for Teaching and Community Development through Action Research: An Overview of the Makerere AgShare Project." *Journal of Asynchronous Learning Networks* 17 (2): 31–42.

Kanwar, Asha, Balasubramanian Kodhandaraman, and Abdurrahman Umar. 2010. "Toward Sustainable Open Education Resources: A Perspective from the Global South." *American Journal of Distance Education* 24 (2): 65–80.

Kemmis, Steven, and Robin McTaggart. 2000. "Participatory Action Research: Communicative Action and the Public Sphere." In *The SAGE Handbook of Qualitative Research*, 2nd ed., edited by Norman K. Denzin and Yvonna S. Lincoln, 567–605. Thousand Oaks, CA: SAGE Publications.

Klijn, Erik-Hans, Jurian Edelenbos, and Bram Steijn. 2010. "Trust in Governance Networks: Its Impacts on Outcomes." *Administration & Society* 42 (2): 193–221.

Loudon, Melissa, and Ulrike Rivett. 2013. "Enacting Openness in ICT4D Research." In *Open Development: Networked Innovations in International Development*, edited by Matthew L. Smith and Katherine M. A. Reilly, 53–78. Cambridge, MA: MIT Press; Ottawa: IDRC. https://idl-bnc-idrc.dspacedirect.org/bitstream/handle/10625/52348/IDL-52348.pdf?sequence=1&isAllowed=y.

Luca, Michael, and Georgios Zervas. 2016. "Fake It Till You Make It: Reputation, Competition, and Yelp Review Fraud." *Management Science* 62 (12): 3412–3427.

Lukić, Tin, Ivana Blešić, Biljana Basarin, B. Ljubica Ivanović, Dragan Milošević, and Dušan Sakulski. 2014. "Predatory and Fake Scientific Journals/Publishers—a Global Outbreak with Rising Trend: A Review." *Geographica Pannonica* 18 (3): 69–81. http://www.dgt.uns.ac.rs/pannonica/papers/volume18_3_3.pdf.

McKnight, D. Harrison, and Norman L. Chervany. 2001. "Trust and Distrust Definitions: One Bite at a Time." In *Trust in Cyber-societies: Integrating Human and Artificial Perspectives*, edited by Rino Falcone, Munindar Singh, and Yao-Hua Tan, 27–54. London: Springer-Verlag.

Middleton, James. 2012. "Operators Can Reinvent Themselves as the Industry's Innovators." *Telecom.com*, September 3. http://telecoms.com/48747/operators-can-reinvent-themselves-as-the-industrys-innovators/.

Morawczynski, Olga, and Gianluca Miscione. 2008. "Examining Trust in Mobile Banking Transactions: The Case of M-Pesa in Kenya." In *Social Dimensions of Information and Communication Technologies Policy: Proceedings of the Eighth International Conference on Human Choice and Computers (HCC8), IFIP TC 9*, Pretoria, South Africa, September 25–26, edited by Chrisanthi Avgerou, Matthew L. Smith, and Peter van den Besselaar, 287–298. New York: Springer.

Mulder, Jorrit. 2008. "Information Imperialism." In *Knowledge Dissemination in Sub-Saharan Africa: What Role for Open Educational Resources (OER)*, 58–67. Amsterdam: University of Amsterdam.

National Institute of Urban Affairs. 2015. *Compendium of Global Good Practices: ICT in Urban Services*. New Delhi: National Institute of Urban Affairs. https://smartnet.niua.org/sites/default/files/resources/GP-GL1_ICT.pdf.

Notenboom, Leo A. 2014. "The Problem with Online Reviews." *Ask Leo!* (blog), February 16. https://askleo.com/the-problem-with-online-reviews/.

O'Hara, Kieron. 2004. *Trust: From Socrates to Spin*. London: Icon Books.

Olleros, F. Xavier. 2008. "Learning to Trust the Crowd: Some Lessons from Wikipedia E-Technologies." In *Proceedings of the International MCETECH Conference on e-Technologies*, Montreal, Quebec, January 1, 2008. IEEE Computer Society. https://papers.ssrn.com/sol3/papers.cfm?abstract_id=2294179.

Ortmann, Jens, Minu Limbu, Doug Wang, and Tomi Kauppinen. 2011. "Crowdsourcing Linked Open Data for Disaster Management." In *Terra Cognita 2011: Foundations, Technologies and Applications of the Geospatial Web, Proceedings of the International MCETECH Conference on E-Technologies*, edited by Rolf Grütter, Dave Kolas, Manolis Koubarakis, and Dieter Pfoser, 11–22. Bonn. http://ceur-ws.org/Vol-798/paper2.pdf.

Ostrom, Elinor, and James Walker. 2003. *Trust and Reciprocity: Interdisciplinary Lessons for Experimental Research*. New York: Russell Sage Foundation.

Peacock, Trevor, Geoff Fellows, and Ken Eustace. 2007. "The Quality and Trust of Wiki Content in a Learning Community." In *ICT: Providing Choices for Learners and Learning*, Proceedings of ascilite 2007, Singapore, Centre for Educational Development, Nanyang Technological University, 2007, edited by R. J. Atkinson, C. McBeath, S. K. Soong, and C. Cheers, 822–832. . http://www.ascilite.org/conferences/singapore07/procs/peacock.pdf.

Peters, Michael A., and Rodrigo G. Britez, eds. 2008. *Open Education and Education for Openness*. Educational Futures: Rethinking Theory and Practice 27. Rotterdam: Sense Publishers.

Raman, Bhuvaneswari. 2012. "The Rhetoric of Transparency and Its Reality: Transparent Territories, Opaque Power and Empowerment." *The Journal of Community Informatics* 8 (2). http://ci-journal.org/index.php/ciej/article/view/866.

Raman, Nithya V. 2012. "Collecting Data in Chennai City and the Limits of Openness." *The Journal of Community Informatics* 8 (2). http://ci-journal.org/index.php/ciej/article/view/877.

Rao, Anuradha. 2014. "Information and Communication Technologies (ICTs) and Civil Society in an 'IT City': Experiences of Civic and Political Engagement in Bangalore." Unpublished PhD diss., National University of Singapore.

Rao, Anuradha, and Mohan J. Dutta. 2016. "Repertoires of Collective Action in an 'IT City': Urban Civil Society Negotiations of Offline and Online Spaces in Bangalore." *Communication Monographs* 82 (2): 221–240.

Resnick, Paul, Ko Kuwabara, Richard Zeckhauser, and Eric Friedman. 2000. "Reputation Systems: Facilitating Trust in Internet Interactions." *Communications of the Association for Computing Machinery* 43 (12): 45–48.

Roberts, Jeffrey A., Il-Horn Hann, and Sandra A. Slaughter. 2006. "Understanding the Motivations, Participation, and Performance of Open Source Software Developers: A Longitudinal Study of the Apache Projects." *Management Science* 52 (7): 984–999.

Rogers, Everett M. 2003. *Diffusion of Innovations*. 5th ed. New York: Simon and Schuster.

Sfakianakis, Stelios, Catherine Chronaki, Franco Chiarugi, and D. Katehakis. 2007. "Reflections on the Role of Open Source in Health Information System Interoperability." In *IMIA Yearbook of Medical Informatics*, edited by A. Geissbuhler, R.Haux, and C.Kulikowski, 51–61.

Sidorkin, Alexander M. 2000. *Toward a Pedagogy of Relation*. Faculty Publications 17. Providence: Rhode Island College. https://digitalcommons.ric.edu/facultypublications/17.

Singh, Parminder J., and Anita Gurumurthy. 2013. "Establishing Public-ness in the Network: New Moorings for Development—a Critique of the Concepts of Openness and Open Development." In *Open Development: Networked Innovations in International Development*, edited by Matthew L. Smith and Katherine M. A. Reilly, 173–196. Cambridge, MA: MIT Press; Ottawa: IDRC. https://idl-bnc-idrc.dspacedirect.org/bitstream/handle/10625/52348/IDL-52348.pdf?sequence=1&isAllowed=y.

Smith, Marshall S. 2013. "Open Educational Resources: Opportunities and Challenges for the Developing World." In *Open Development: Networked Innovations in International Development*, edited by Matthew L. Smith and Katherine M. A. Reilly, 129–170. Cambridge, MA: MIT Press; Ottawa: IDRC. https://idl-bnc-idrc.dspacedirect.org/bitstream/handle/10625/52348/IDL-52348.pdf?sequence=1&isAllowed=y.

Smith, Matthew. 2007. "Confianza a La Chilena: A Comparative Study of How E-Services Influence Public Sector Institutional Trustworthiness and Trust." Unpublished PhD diss., London School of Economics and Political Science.

Smith, Matthew L., and Ruhiya Seward. 2017. "Openness as Social Praxis." *First Monday* 22 (4). https://firstmonday.org/ojs/index.php/fm/article/view/7073.

Song, Steve. 2015. "The Future of Open and How to Stop It." *Many Possibilities* (blog), January 30. https://manypossibilities.net/2015/01/the-future-of-open-and-how-to-stop-it/.

Srinivasan, Janaki. 2007. "The Role of Trustworthiness in Information Service Usage: The Case of Parry Information Kiosks, Tamil Nadu, India." In *Proceedings of the 2nd IEEE/ACM International Conference on Information and Communication Technologies and Development*, December 15–16, 2007, Bangalore, India, 345–352. Bangalore: IEEE.

Starbird, Kate, Jim Maddock, Mania Orand, Peg Achterman, and Robert M. Mason. 2014. "Rumors, False Flags, and Digital Vigilantes: Misinformation on Twitter after the 2013 Boston Marathon Bombing." In *iConference 2014 Proceedings*, 654–662. Granville: iSchools. http://hdl.handle.net/2142/47257.

Steinmacher, Igor, Marco A. G. Silva, and Marco A. Gerosa. 2014. "Barriers Faced by Newcomers to Open Source Projects: A Systematic Review." In *10th IFIP International Conference on Open Source Systems (OSS)*, May 2014, San José, Costa Rica, IFIP Advances in Information and Communication Technology 427, edited by Luis Corral, Alberto Sillitti, Giancarlo Succi, Jelena Vlasenko, and Anthony I. Wasserman, 153–163. Berlin: Springer.

Sztompka, Piotr. 1999. *Trust: A Sociological Theory*. Cambridge: Cambridge University Press.

Taddeo, Mariarosaria 2010. "Trust in Technology: A Distinctive and a Problematic Relation." *Knowledge, Technology & Policy* 23 (3–4): 283–286.

Tandoc, Edson, Jr., Oscar Westlund, Andrew Duffy, Richard Ling, Debbie Goh, and Lim-Zheng Wei. 2018. "Audience's Acts of Authentication in the Age of Fake News." *New Media & Society* 20 (8): 2745–2763.

Thomas, W. I., and Dorothy S. Thomas. 1928. *The Child in America: Behavior Problems and Programs*. New York: Knopf.

Tomberg, Vladimir. 2013. "Learning Flow Management and Teacher Control in Online Personal Learning Environments." PhD diss., Tallinn University.

Turilli, Matteo, Antonino Vaccaro, and Mariarosario Taddeo. 2010. "The Case of Online Trust." *Knowledge, Technology & Policy* 23 (3–4): 333–345.

Tuttle, Brad. 2014. "5 Outrageous Ways People Try to Game Online Reviews." *Money*, August 6.

UNESCO. 2012. *The Paris OER Declaration*. Paris: UNESCO. https://en.unesco.org/oer/paris-declaration.

Weller, Martin, Bea de los Arcos, Rob Farrow, Beck Pitt, and Patrick McAndrew. 2015. "The Impact of OER on Teaching and Learning Practice." *Open Praxis* 7 (4): 351–361. http://oro.open.ac.uk/44963/1/227-1106-2-PB-3.pdf.

Williams, Sarah, Adam White, Peter Waiganjo, Daniel Orwa, and Jacqueline Klopp. 2015. "The Digital Matatu Project: Using Cell Phones to Create Open Source Data for Nairobi's Semi-formal Bus System." *Journal of Transport Geography* 49:39–51.

World Bank. 2013a. *Philippines—Transport Crowd-source ICT Demonstration: Final Report.* Washington, DC: World Bank. http://documents.worldbank.org/curated/en/145751468092669224/Philippines-Transport-crowd-source-ICT-demonstration-final-report.

World Bank. 2013b. "Mexico City Open Database Improves Transit Efficiency, Helps Commuters." World Bank, November 5. http://www.worldbank.org/en/news/feature/2013/11/05/mexico-city-open-database-improves-transit-efficiency-helps-commuters.

Zaheer, Srilata, and Akbar Zaheer. 2006. "Trust across Borders." *Journal of International Business Studies* 37 (1): 21–29.

4 Learning as Participation: Open Practices and the Production of Identities

Bidisha Chaudhuri, Janaki Srinivasan, and Onkar Hoysala

Introduction

For some time, the world has been looking hopefully toward digitally enabled *openness* to bring about positive transformation and development (Smith, Elder, and Emdon 2011). In this chapter, we unpack this hope and examine the linkages between open initiatives and development. The prefix *open* conjures up the idea of making digital platforms, knowledge, and knowledge development processes more accessible, including to a hitherto excluded group of people. However, the links between openness, participation, and development are far from automatic, and understanding people's participation in open processes continues to elude researchers and practitioners. Thus, what we need to focus on is not only whether participation occurred but *who* participated and *who* was excluded (whether by exercising their choice or systematically). In sum, there is a need to understand how existing micro and institutional power structures shape the dynamics of participation. Moreover, open development cannot afford to focus merely on the outcomes of an intervention and label them a success or failure relative to the goals of that intervention. We need to focus equally on the processes and practices[1] by which those outcomes were reached.

This chapter develops a framework for a better understanding and analysis of open development processes that link people's participation in open practices to open development outcomes through changes in their identities. Understanding people's participation in open processes involves analyzing "what kind of participation and to what avail, on whose terms it takes place, and how it recasts power" (Singh and Gurumurthy 2013, 176). Harvey (2013, 284) likewise pointed out in his study of AfricaAdapt that participation in a collaborative learning network is contingent on "the types of tools or resources made available for users to participate..., the forms of invitation they receive to participate, the incentives for or pressures to accommodate particular actors over others [and so on]." Different tools, skills, and opportunities are needed to engage in open practices, but it is still not clear how individuals learn the required skills or how to use new tools in order to participate, particularly within informal contexts. This framework provides a way to understand the different types of learning taking place within open processes and the elements of change in the identities of the individuals that emerge from it.

The framework has two key elements. First, it draws on Smith and Seward's (2017) framework of open practices to map out a research agenda for this purpose. Second, it builds on the idea of learning as participation, taken from situated learning theory. We argue that *learning* is more than just building the skills to navigate the components of a particular open initiative. Furthermore, open practices do more than merely enhance or limit what actors can or cannot do. Instead, we argue that *learning as participation* in open practices can fundamentally shape actors' identities. Their evolving identities, in turn, provide them with opportunities to change how they lead their everyday lives beyond the open initiative. Our focus on *learning as participation* allows us to identify change as hinging on an individual's social and cultural context, keeping relations of power central to our framework. If *learning as participation* is the main modality through which people engage with open processes, it may therefore be the central way by which social transformation happens. It is therefore crucial to understand how individuals are learning to participate and how such participation changes their identities.

This chapter starts by briefly outlining the core concepts of situated learning theory, followed by a framework linking open practices and learning as participation. It then builds on situated learning theory by contributing the notions of instrumental versus substantive learning, arguing that substantive learning—learning that shapes an actor's identity—is an important outcome for open practices to strive for. It concludes by outlining a research agenda for connecting learning as participation, open practices, and the production of identities.

Situated Learning Theory and Open Practices

Our framework builds on Smith and Seward's (2017) work on open practices that span production, distribution, and consumption in an open initiative.

These include peer production, crowdsourcing, sharing, republishing, remixing, retaining, and reusing of content. Smith and Seward's framework views these practices as less or more open, rather than as open or closed. In building our framework linking open practices and learning as participation, we keep to this understanding of openness. This chapter explores how people learn to participate in these open practices and the subsequent implications for individual and social transformation.

The starting point is to view open practices as a type of social participation. However, everyone who engages with or encounters an open development initiative experiences change differently and is able to make sense of these changes from their own social, cultural, and historical positions. How people *learn* to make sense of and cope with the contextualized changes induced by participating in open practices is itself a social transformation.

This understanding of learning draws on situated learning theory, which defines learning as "a social phenomenon constituted in the experienced, lived in world, through legitimate peripheral participation in ongoing social practice" (Lave 1991, 64). Such a treatment of learning—as constituted through participation—sits particularly well with open initiatives, with their emphasis on increased participation. In terms of the mechanisms of learning and change, we also focus on how learning goes on to mold the identities of those engaged in open initiatives and influences their everyday lives as a key outcome of participating in open practices. The following two subsections contextualize this way of understanding learning and briefly outline situated learning processes.

How Situated Learning Theory Applies to Open Practices

Situated learning theories view learning as a social process. They emphasize that learning is always situated within culturally organized settings (Lave 1988; Talja 2010). Situated theories of learning allow us to address how learning takes place in a variety of social situations outside formal, structured environments of learning (Lave 1988; Lave and Wenger 1991; Lave 2011; Wenger 1998).[2] Here, learning is not seen as "a separate activity, it is not something we do when we do nothing else or stop doing when we do something else" (Wenger 1998, 5; Wenger, White, and Smith 2009). It is different from being taught (Lave and Wenger 1991). Learning in this framing takes place not within an individual but in a cultural historical setting of a community of practice (CoP).

Situated learning theories claim that "every human thought is adapted to the environment, that is *situated*, because what people *perceive*, how they *conceive of their activity*, and what they *physically do* develop together" (Clancey 1997, 1, emphasis original). This allows us to address how learning takes place in a variety of social situations (Lave 1988; Lave and Wenger 1991; Lave 2011; Wenger 1998). In order to reflect typical processes of engaging with open practices, we focus primarily on settings where learning is not the primary goal, but these theories still apply to open practices within any setting.

There are two key aspects of situated learning theory that are important for understanding how learning takes place within open practices. First, knowledge is conceptualized as lived practice or a "product of the activity, context and culture in which it is developed and used" (Brown, Collins, and Duguid 1989, 32). By thinking of knowledge as a tool, Brown, Collins, and Duguid (1989, 33) distinguish between "the mere acquisition of inert concepts and development of useful, robust knowledge" and "learning how to use a tool [that is, knowledge, which] involves far more than can be accounted for in any set of explicit rules. The occasions and conditions for use arise directly out of the context of activities of each community that uses the tool, framed by the way members of that community see the world." This fundamentally changes the focus of learning from individuals or general outcomes to specific social and cultural settings and the practices of people within them. It is therefore impossible to understand an individual's reasons for sharing an open resource until we understand whether sharing is common or typical within their particular culture or a result of the specific task at hand and how underlying power relations shape the activity of sharing. Situated learning theory insists that learning is rooted in sociocultural settings and requires communities of practice. Thus, this theory enables us to focus on the sociocultural contexts within which open practices take place.

The second key aspect is that learning is conceptualized as increasing participation in communities of practice. To grasp this concept, consider that communities of practice represent the intertwined nature of individuals, their relationships, and their actions within a sociocultural community. Communities of practice are "formed by people who engage in a process of collective learning in a shared domain of human endeavour" (Wenger-Trayner and Wenger-Trayner 2015). Increasing participation in a CoP means that individuals learn by interacting and engaging more with the sociocultural community over time by gradually adopting a shared repertoire of resources and eventually developing shared histories of experience.

Furthermore, learning as participation in communities of practice draws attention to the ways in which individuals develop their own learning trajectories[3] in multiple communities of practice (Wenger 1998).

Situated learning theories offer three useful insights for an analysis of open initiatives. First, they help explain how learning takes place in informal or nonformal settings where learning might not be the primary goal. This helps us capture the informal learning that happens when participating in open practices.

Second, situated learning theory invokes the link between participation and identity within an open initiative. For instance, consider an open initiative where community members are invited to a public consultation with development project managers and government officials. They might be solicited for their opinions, which are then synthesized and published publicly online. Through attending a meeting, community members could gain a sense of participation by interacting with project managers and government officials. This kind of open practice is consistent with a CoP that reflects traditional development roles, such that community members are contributors and project managers are decision makers who maintain ultimate control over what information to consolidate and share publicly. In contrast, if the open initiative instead crowdsources development project ideas and permits community members to vote on the most important initiative, the members enter into a CoP that fundamentally changes the roles of both the community members and project managers. This indicates how such learning "shapes not only what we do, but also who we are and how we interpret what we do" (Wenger 1998, 4).

This brings us to the third insight, which is that invoking the link between participation and identity implies carefully examining power relations in context. Practitioners often gloss over existing power structures and how they influence who can use open initiatives in practice (Davies and Bawa 2012; Singh and Gurumurthy 2013; Srinivasan 2011). How different perceptions are built by the same initiative, how certain voices get more space, and how particular interests and alliances that are forged denote different paths of participation (Harvey 2013). If the process through which people learn to make sense of an open initiative shapes the identities of users and how they lead their everyday lives, then it is necessary to recognize how and why power differentials influence paths of participation that take root.

Situated Learning Processes

We now examine the specifics of situated learning processes. These will then be contextualized and expanded further in the next section. We draw primarily on Lave and Wenger's (1991) legitimate peripheral participation, or *learning as participation*, because it explains how an individual learns to participate in social and cultural practices and how they develop the capabilities to do so.

Lave and Wenger's (1991) process of legitimate peripheral participation concentrates on the ways in which people develop a sense of belonging to a CoP. The use of the term *legitimate* reflects how people gain access to communal resources and the opportunity to control them. Peripheral participation describes differences between those who are newcomers to a CoP and those who are considered full participants that belong. Over time, newcomers can move centripetally toward more intensive participation and to the center of the community to become full participants. This change in location and perspective in moving from peripheral to central or core participation is part of actors' personal learning trajectories, the development of their identities, and forms of membership (Lave and Wenger 1991). Legitimate peripheral participation is therefore a complex process that is "implicated in social structures involving relations of power" (Lave and Wenger 1991, 36). This focus on power relations within *learning as participation* contributes a much-needed layer of analysis within open development.

Furthermore, as learning consists of activity, concepts, and culture (Brown, Collins, and Duguid 1989), cognitive tasks are never carried out solely inside the head. Instead, these tasks are solved by constantly drawing on the environment, which includes social settings as well as physical infrastructure and artifacts. In this sense, learning and doing are inseparable. Consider as an illustration a community that uses a public Facebook group to share knowledge surrounding community cleanup and recycling initiatives. In this scenario, members must learn how to access the Facebook group and apply the shared knowledge to their own situations. At one point, a new member decides to attend a cleanup event and is able to meet various people who have come to clean up, all having their own methods and reasons for cleaning up. Later, one of the event organizers asks the member to share what he has done with the community. The member has never participated actively in the Facebook group before, but since Facebook was his only prior source of interaction with the community and because he had seen other members sharing photos and quotations from members at events, he suddenly knew to begin taking photos and sharing different members' perspectives with the group. Here, the new member is learning to share within the community by drawing on both the social norms of the Facebook group and the material platform that Facebook offers. Thus, we see that "the occasions and conditions for use [in this case, taking and sharing pictures of the event] arise directly out of the context of activities [of the Facebook group, in this case]...framed by the way members of that community see the world" (Brown, Collins, and Duguid 1989, 33).

It is important to note that *learning as participation*—the situated learning concept that frames legitimate peripheral participation—has been conceptualized according to many levels of analysis, including those of individuals, communities, and organizations (Wenger 1998). For instance, researchers and practitioners may intuitively wish to define a CoP in the context of open initiatives for relevant study. However, our focus is on using situated learning theory to understand the co-constitutive relationships between individuals, open practices, power relations, and contexts rather than on identifying characteristics and outcomes of a CoP where it may exist. This distinction is also useful for the study of open practices since we cannot presuppose the physical copresence of users of open initiatives. Users do learn how to navigate the open initiative for their purposes and may participate in multiple communities of practice simultaneously, so the questions we might raise by focusing on learning as increasing participation in a CoP include: What sorts of resources or communities of practice do they form and draw on for this learning? And what kinds of organizational forms, as studied by Mateos-Garcia and Steinmueller (2008), and what spatial or relational proximity, as studied by Amin and Roberts (2008), enable this learning?

Building a Theoretical Framework for Connecting Learning as Participation, Identity, and Open Practices

Learning as participation shapes one's identity and therefore one's agency and the negotiation of life situations. The capacity to participate in open practices can bring about positive social transformation for individuals. There are two main parts to our framework. The first expands on situated learning processes to clarify the type of learning taking place within open practices. The second then focuses on how participation in open practices potentially shapes one's identity.

Clarifying the Type of Learning Taking Place: Instrumental and Substantive Learning

In this section, we draw on the work of our SIRCA III (Strengthening Information Society Research Capacity Alliance) colleagues who applied our framework in their empirical research in order to iteratively reformulate dimensions of learning. This empirical work involved the ethnographic study of an open initiative that was part of the Adaptation Fund.[4] Part of a multiyear effort to support climate change adaptation, the initiative produced and disseminated weather forecasts and associated agricultural recommendations to small-scale and marginal farmers in two districts in West Bengal in India using a variety of channels, including blackboards, weather bulletins, and a short message service (SMS) (see Kendall and Dasgupta, chapter 7, this volume). It sought to enable farmers to better understand weather patterns and appropriately respond by changing their agricultural practices. To achieve its vision, the initiative introduced a range of technologies (rain gauges and other weather-related instruments, as well as traditional and digital information communications technologies); processes for producing, disseminating, and consuming weather and agriculture-related content; and a network of practitioners that included experts and novices. The SIRCA III empirical study examined the open practices of this initiative to understand how different categories of actors involved in this open initiative experienced and learned in practice and what kinds of identities it helped them foster.

In mapping diverse open practices to learning as participation, we propose two dimensions of learning: instrumental and substantive. These dimensions build on situated learning processes to clarify the types of learning that are critical to identity transformations. By instrumental learning, we mean the learning of techniques and skills, in this instance through engaging in an open initiative. These skills can be deployed for specific uses as prescribed by the tools and procedures set out by the initiative or one can deploy these skills to navigate beyond a particular open initiative. For example, in the West Bengal case study, the production of weather information for farmers involved active participation by field executives in different villages. In the course of producing these weather bulletins, these field executives acquired many skills, including reading manual rain gauges and other tools as well as translating weather advisories from English to Bengali.

In contrast, substantive learning refers to learning that shapes aspects of the user's identity and ability to navigate socially over the long term. In the same initiative, our empirical team found that while the initiative was distributing information for agricultural purposes, several villagers were in fact using this information to decide whether to send their wards to school, and bricklayers were using these predictions for making crucial decisions in their day-to-day work practices. Furthermore, in the course of producing weather bulletins and crop advisories, the field executives frequently ended up interacting with a meteorologist and agricultural scientists on the one hand and farmers and village weather kiosk staff on the other. In the process, the field executives not only acquired new skills but also built up their knowledge base related to meteorology and agricultural sciences to a level where they were considered experts by the local population. Being perceived as experts shaped the executives' identities in a substantive way, allowing them to leverage their newfound status within the open initiative.

We do not consider the shift from instrumental to substantive learning a sequential one, nor do we consider it inevitable. Take the example of the field executives. We can imagine a situation where a less skilled field executive is recognized as an expert while a more skilled one fails to be recognized as such. Thus, we find that learning is a layered process with instrumental and substantive aspects that are important to discern, albeit difficult to separate. The distinction between them is purely for analytical purposes; it might even be hard to distinguish between them immediately upon observation. That said, for us, the crucial distinction between the two is that while the instrumental aspects of learning are tied to skills to enable the use of tools in an open initiative, the substantive aspects of learning are closely tied to questions of participation and identity. However, it is important to keep in mind that substantive learning does not always imply positive social transformation. This is why the substantive aspects of learning will always appear *in relation to* increasing participation in a CoP, which is inherently governed by power relations in context.

While we do not dismiss instrumental learning, for three reasons we find it critical to go beyond it to understand substantive learning in the context of open practices. First, learning to use common open practice tools, such as mobile phones, or how crowdsourcing works in the context of one open initiative may lead to improved use in another. However, it is when users practice applying these tools in different settings, developing personal learning trajectories across multiple communities of practice, that they may be capable of deriving value from open practices more substantively. Substantive learning helps the user figure out how to practice those skills in ways that help them negotiate important life situations. Second, since it works at the level of a user's identity, substantive learning need not be applicable only to a user's engagement in one kind of open practice. It can instead influence a user's actions and abilities in a range of domains that are beyond the boundaries of an open initiative. Consequently, and finally, it also allows for the possibility that substantive learning will shape further development processes that may not even be directly associated with digital platforms and systems. We acknowledge that gauging substantive learning requires a long-term understanding of a person's social position, the complexity of the contexts that she or he inhabits, and also her or his everyday open practices. In the following discussion, we flesh out how researchers might approach this task.

Understanding How Open Practices Shape One's Identity

As our point of departure, we explore the following questions about learning as participation in the context of openness:

1. How does engaging in the three types of open processes shape instrumental and substantive learning?
2. How and why do the ways in which open practices shape learning influence one's identity?

In order to tackle these questions, we begin by elaborating on the linkages that table 4.1 presents. Consider, for example, the practice of peer production as a type of open production. Over time, a person participating in this process may develop an ability to recognize whether information is of *good* enough quality to add to a collective repository. This might constitute instrumental learning for that platform. Meanwhile, after repeated additions of good-quality data or pointing out bad-quality data, a participant might start being identified as a full participant in a subcommunity of that peer production platform. In turn, this might shape how a participant regards herself within that community and fundamentally shape her identity in this process. This relationship between substantive learning and increasing

Table 4.1
The three main types of open processes, key characteristics of open practices, and learning as participation

Open process	Open practices	What constitutes learning in this practice
Open production	Peer production	*Instrumental learning* Acquiring new skills to create content, to identify good quality and range of information, entering new networks of collaboration with implications for career opportunities *Substantive learning* Becoming, and being seen as, an expert in producing relevant content, increasing participation in a CoP, being more respected within one's immediate affective community of friends and family
	Crowdsourcing	*Instrumental learning* Learning to identify good quality and range of information, skills to navigate and create content *Substantive learning* Increasing participation in a community of experts, being more respected within one's immediate affective community of friends and family, may lead to greater sense of participation across multiple communities, deeper sense of belonging
Open distribution	Sharing, republish	*Instrumental learning* Skills to identify relevant community groups and various ways to share and republish information with them *Substantive learning* Becoming, and being seen as, an expert in sharing and republishing information by a larger community, forming wider social networks
Open consumption	Retain, reuse, revise, remix	*Instrumental learning* Skills to discern and gauge the quality of information, to identify what information to consume and how to consume it *Substantive learning* Seeing *themselves* as experts in dealing with digitized information and discerning its quality, being seen as an expert within a group of consumers, leading to an improved position within it

Note: An expert is someone who is considered as such by the community. An expert in one aspect of practice within a community need not be an expert in all other aspects of the practices of the community. A full participant in a CoP is anyone who legitimately engages with the CoP through shared resources and shared histories of the practice (Lave and Wenger 1991; Wenger 1998). A full participant may be a peripheral participant or a central one.

participation in communities of practice may be reflected especially in the voluntary nature of these practices. For instance, updating traffic information on the Twitter handle of the civic authorities may cement people's feeling of contributing to the larger community and also enhance contributors' sense of belonging when they are acknowledged by other members of the community. In table 4.1, we make a few more such suggestions of linkages between open practices and learning.

However, focusing on the type of learning one might expect to see within a particular open process may be insufficient for understanding identity transformations when in fact understanding substantive learning and the accompanying shifts in identity rely also on the social and cultural settings surrounding open practices. It is also important to keep in mind a broader understanding of *learning as participation* that goes beyond our analytical separation between instrumental and substantive learning. Our theoretical framework primarily addresses ways to understand learning as mediated by the experiences of individuals as they make sense of tools and practices from their social, cultural, and historical positions. These different meaning-making processes and their subsequent incorporation into individuals' lives are what form learning as participation. These types of learning can be studied through a nuanced understanding of their everyday practices and in relation to personal learning trajectories within a CoP.

We emphasize that the extent and type of participation in a CoP are shaped by where and how open processes take place,[5] the domain within which an information system is embedded, and the social location of a potential participant.[6] For example, in the case of the weather information system in West Bengal, open production practices may enable village volunteers to learn climate change mitigation practices important to the community. This may then lead to substantive learning if these volunteers increasingly participate as experts in this domain. However, the participation of individuals is itself constrained and shaped by their social positions. For example, an individual's gender, caste, class, religious identity, or literacy level may play a role in the kinds of open production, distribution, or consumption practices they are able to engage in. In the case of women farmers in the same West Bengal case, their chances of engaging in open production were much lower. They may face obstacles to participating or collaborating in open practices because of their existing skill levels, the cultural settings that dictate where men and women work and socialize, or the communities that it is socially acceptable for them to work with on a regular basis. Here, investigating how people increasingly participate in open practices encourages researchers to consider both those who are fully participating and those who may be excluded or be on the periphery. For example, we can see how learning actually takes place among volunteers by analyzing everyday interactions between old and new volunteers. We can see how farmers are able to move across open practices—for example, from open consumption to open distribution—by developing a sense of belonging to the CoP that governs a particular open process in context.

Let us take another example of open production—peer production of open source software. A newcomer willing to contribute code typically begins as a contributor who *commits* code to the repository. This code is reviewed by moderators and, if approved, is *merged* into the main code base. The instrumental learning in this situation is to learn how to use the tools to undertake the practices of committing code to a code base and understanding what the community interprets as adherence to their standards of code. Different open source projects have different project structures,[7] so, even if developers are not physically located within the same cultural setting, there is still a culture that develops around the initiative that is important to consider. Based on the kind of project, a contributor could, over time, belong to the community, learn the nuances of the project, and be considered an expert at it. Legitimate peripheral participation implies that people begin to participate peripherally and then move to a more central role over time through interactions with the community. In contrast, we emphasize that it is equally important to consider the personal learning trajectories of individuals who may be at a disadvantage because of their social location. The social location of individuals must be considered not only from within the open initiative (that is, the open source software community) but also from within their personal cultural setting (that is, the circumstances that enable a person to participate in the initiative to begin with). Understanding how people negotiate their social locations via their personal learning trajectory, or not, implies developing a rich understanding of power relations in context.

Similarly, it would also be useful to explore which of the open processes or practices are amenable to leveraging existing communities of practice within a cultural setting. As we mentioned earlier, communities of practice exist prior to the introduction of an open initiative and its open practices. Drawing once more on the case of the weather information system in West Bengal, Kendall and Dasgupta (see chapter 7, this volume) found that the community of farmers were already engaged in a CoP in that they had shared histories of farming practices and were part of a shared domain of human endeavor (Wenger-Trayner and Wenger-Trayner 2015). Conversely, it may be that certain open processes are restricted or taboo because of preexisting cultural practices. In either case, the cultural and social settings surrounding open practices offer us fertile ground and a site for understanding how social relations and power structures shift, are maintained, or are morphed upon the introduction of open practices.

In the previous two paragraphs, spatial versus relational proximity became a major source of debate and contention for connecting learning, participation, and the formation of identities. While one group of scholars recognizes both local and distributed spatial dimensions of communities of practice, others still emphasize the importance of copresence (Amin and Roberts 2008). However, there is a growing consensus among scholars of situated learning that both spatial and relational dimensions are crucial and come in different configurations (Amin and Roberts 2008). One factor that we find to be particularly relevant to open practices is the multiple modes and media through which participants interact, which requires a multidimensional understanding of spatial and relational proximity. For instance, in the case of the farmers using the weather information system in West Bengal, the organization responsible for introducing the information system designed an SMS system in two languages. The idea was to share weather information over SMS with farmers who had phones and who had registered their phone numbers with the organization. These farmers would then share this information with other farmers either through SMS or by other means (including verbally and in person). Multiple practices develop that impact the different ways in which people participate. To take the example of open source software production, many interactions between people are mediated through pieces of software called version control systems or continuous integration systems. Various questions are raised by these practices, such as "What does such mediation hold for how people participate?" and "How do we analytically distinguish between interactions that are purely digital and those that are mediated digitally but occur in the same physical space and with participants having in-person interactions?"

A second challenge we encountered was that, while we distinguish between the three categories of open processes in our framing, the actors involved with open practices on the ground were not always distinct (as also discussed by Smith and Seward 2017). Thus, the empirical team found users of the system who were not only the consumers of information but also its distributors. To follow other examples we have discussed in this chapter, open source enthusiasts are often both the developers and consumers of their software (Weber 2004). On crowdsourcing platforms such as Ushahidi, the producers of information may also be its distributors, if the Ushahidi platform owners set up the software that way. It is worth exploring both theoretically and empirically whether this overlap might prove productive for instrumental and substantive learning. This is also why we argue that focusing on the personal learning trajectories of individuals is a key way to understand the impacts of open practices since individuals participate in multiple communities of practice.

A Research Agenda for Understanding How Participation in Open Practices Shapes Identity

Our framework is rooted in an understanding of learning that is situated in a specific history, geography, and constellation of social relations. Such an understanding of learning will make a priori identification of its specific aspects that will be universally applicable across contexts both difficult and redundant. Hence, in the present formulation, we leave it to empirical research (ideally based on ethnographic research) to provide a detailed understanding of how and why open practices shape identity. Empirical studies that will use our theoretical lens in the field may find practices that are not captured within our model but inform our theory nonetheless. We are open to such findings that can then broaden our perspective, locate gaps in the model, and point toward new dimensions of open practices and learning. We believe that this inductive approach will make our framework more inclusive. In the following discussion, we summarize key questions researchers and practitioners might consider in future investigations of learning as participation in open practices.

An investigation of learning as participation does not assume that individuals or communities interpret open processes as we have laid them out. Even though digital tools may be more commonly associated with open processes, in the West Bengal case for instance, a dominant practice was to use a blackboard, which is not a digital artifact, to share information publicly. While engaging with a specific context, it is necessary for researchers to conceptualize open practices as practices that community members have an opportunity to observe, participate in, and develop shared understandings about. For example, communities may not know what *crowdsourcing* means, but they may have developed their own terminology or collective wisdom around a similar practice. Hence, it is critical for researchers to approach sociocultural contexts with an appreciation for the interpretive flexibility of communities regarding openness.

How and Why Do Communities of Practice Form around Open Practices?

The next step is to identify whether and how communities of practice form around open processes. This involves identifying key groups of actors, how different actors engage in open practices, and to what ends. Remember that spatial proximity might be critical for the formation of communities of practice in some instances (see our example from West Bengal about the farmers' CoP with and without the open initiative), but it may not affect the formation of others (think again about open source or crowdsourcing communities). What is worth considering is whether distance reinforces or dilutes the three dimensions that define communities of practice: their interaction, their shared repertoire of resources, and their shared histories of learning.

How Do Existing Micro and Institutional Power Structures Shape the Dynamics of Participation in Open Practices, and Why?

It follows that if some groups of actors have significantly more or fewer opportunities to observe and engage in open practices, researchers will need to uncover the social and cultural reasons for this. For instance, do open practices enable individuals to increase their participation in communities of practice that were hitherto closed to them? What existing power relations are playing out in context? For instance, are communities of practice formed primarily of established peers that maintain strong social bonds? What are the barriers in expanding their notion of peers? How are newcomers treated? What learning trajectories are afforded to new entrants, and how do newcomers increase their participation in an open practice? Regarding preexisting sociocultural influences specifically, what preexisting practices are used within the community, and how do they differ from those formed around an open practice? For example, if new communities of practice have been formed around distribution, what is their relationship with older channels and practices involved in distribution? Are the new channels and practices replacing, co-opting, or working with them? Does this bring the possibility of conflict and a change in the nature of the mutual engagement at the core of communities of practice?

How and Why Do Dominant Open Practices Aid or Obstruct Substantive Learning?

There may be multiple and/or overlapping communities of practice around an open initiative. Becoming a full participant in one aspect of an open initiative may be available to some people, while full membership may mean significantly fewer opportunities for substantive learning within another aspect of the open initiative. For example, substantive learning opportunities differ significantly for users and producers of open source software, so the questions of interest become: Are there specific open practices that lead to the formation of different kinds of communities of practice? If so, do these practices create synergies or conflicts? In the case of conflict(s), how is a resolution reached (or not)? What implications may such conflict(s) have for substantive learning? If there are communities of practice that existed before the introduction of an open initiative, how do they engage with new ones, if at all? Given that communities of practice may emerge through open practices, how does increasing participation shape substantive learning in both anticipated and unintended ways? Answering these questions will take researchers to underlying relations of power and its link to participation. To understand this, ethnographic work has to focus on the interactions between people and information and what resources people use, juxtaposing them with history.

Who Does Increasing Participation in a Community of Practice around Open Processes *Benefit,* and *Why?*

Understanding dominant practices, sociocultural influences, and how these enable substantive learning is critical for drawing conclusions about shifts to identity on a case-by-case basis. Additionally, it is crucial to construct a broader understanding of these shifts within community contexts. This helps in connecting whether shifts in identity resulting from open practices are one-off occurrences or more structural in nature. This line of research therefore aims to uncover power relations in context and inspires new questions for consideration. For instance, are new communities of practice built around open practices that rearrange power relations of the earlier ones? Do participants stemming from both peripheral and central locations within a CoP have the same voice within an open initiative? Are there barriers to participation for a specific group of actors? Are there strategies in place to mediate power differentials and barriers to participation?

Conclusion

Our theoretical framework discusses learning as a continuous process with both instrumental and substantive dimensions. We also propose that situated learning theory offers us an analytical lens for understanding how learning happens within open practices. It is important to mention here that we do not presuppose that substantive learning has only positive implications for open development. As we see substantive learning as social activity tied to the notion of identity, it contains different political possibilities for different groups of people. Hence, we argue that positive and negative impacts of substantive learning will be contingent on a plethora of other factors that determine the axes of inclusion and exclusion. The power and understanding of substantive learning will emerge in relation to the extent to which individuals increase participation in communities of practice surrounding open processes. We suggest that ethnographic studies that probe the nuances of participation in open practices will be essential for understanding how open practices enable such learning for diverse social groups.

In creating our framework, we distinguish between theoretical generalization and scaling across social contexts; our goal here is the former. We therefore anticipate that empirical research will take our framework as a point of departure and then draw out those details of indicators required to analyze open practices around information systems, and their learning dimensions, that are best suited to the domains and contexts a specific empirical study chooses to examine. In this chapter, however, we provided a few examples for illustrative purposes, with added emphasis on the West Bengal case. We prioritized this case to illustrate our framework because of its direct influence in reformulating that framework.

While its potential is widely accepted, open development necessarily challenges existing power structures and the status quo (Reilly and Smith 2013; Smith, Elder, and Emdon 2011). This may not only lead to "disruptive transformation" but also implies that open development constitutes spaces of constant struggle (Smith, Elder, and Emdon 2011, v–vi). With our framework, it is this shifting space and its implications for development that we hope to map using studies of how learning is negotiated by diverse participation in open practices in various social circumstances.

Notes

1. Brown and Duguid (2000, 95) discuss practices as the internal life of processes. They note, however, that processes within workplaces are often codified, and tensions exist between "the demands of processes and the needs of practice."

2. In critiquing learning-transfer theories, Lave draws from theories of practice (Bourdieu 1972; Giddens 1984) to build on the idea that "everyday activity is…a more powerful source of socialization than intentional pedagogy" (Lave 1988, 14).

3. We intend to situate identity construction in multiple social contexts and to consider the fluidity of how individuals negotiate different environments and construct their own identities amid this mobility. When we refer to individual or personal learning, we are not referring to merely psychological dimensions of identity.

4 The Adaptation Fund was set up through Kyoto Protocol of the United Nations Framework Convention on Climate Change (UNFCCC) as an international fund to finance climate change adaptation and mitigation.

5. For example, much research in information and communications technologies for development suggests that what people do with ICTs is shaped heavily by whether these are accessed by individuals or by groups, in public community centers or in individual homes, and the devices on which information systems are accessed.

6. Whether the information system in question is in the domain of health, education, or governance significantly shapes its technological structures and the social groups who access it.

7. As well as other associated power structures. For example, Apache and Linux have vastly different power structures.

References

Amin, Ash, and Joanne Roberts. 2008. "The Resurgence of Community in Economic Thought and Practice." In *Community, Economic Creativity, and Organization*, edited by Ash Amin and Joanne Roberts, 11–34. Oxford: Oxford University Press.

Bourdieu, Paul. 1972. *Outline of a Theory of Practice*, edited by Ernest Gellner, Jack Goody, Stephen Gudeman, Michael Herzfeld, and Jonathan Parry. Cambridge: Cambridge University Press.

Brown, John S., Allan Collins, and Paul Duguid. 1989. "Situated Cognition and the Culture of Learning." *Educational Researcher* 18 (1): 32–42.

Brown, John Seely, and Paul Duguid. 2000. *The Social Life of Information*. Boston: Harvard Business School Publishing.

Clancey, William J. 1997. *Situated Cognition: On Human Knowledge and Computer Representations*. New York: Cambridge University Press.

Davies, Tim G., and Zainab A. Bawa. 2012. "The Promises and Perils of Open Government Data (OGD)." *The Journal of Community Informatics* 8(2).

Giddens, Anthony. 1984. *The Constitution of Society: Outline of the Theory of Structuration*. Oakland: University of California Press.

Harvey, Blane. 2013. "Negotiating Openness across Science, ICTs, and Participatory Development: Lessons from the AfricaAdapt Network." In *Open Development: Networked Innovations in International Development*, edited by Matthew L. Smith and Katherine M. A. Reilly, 275–296. Cambridge, MA: MIT Press; Ottawa: IDRC. https://idl-bnc-idrc.dspacedirect.org/bitstream/handle/10625/52348/IDL-52348.pdf?sequence=1&isAllowed=y.

Lave, Jean. 1988. *Cognition in Practice: Mind, Mathematics and Culture in Everyday Life*. New York: Cambridge University Press.

Lave, Jean. 1991. "Situating Learning in Communities of Practice." In *Perspectives on Socially Shared Cognition*, edited by Lauren B. Resnick, John M. Levine, and Stephanie Teasley, 63–82. Washington, DC: American Psychological Association.

Lave, Jean. 2011. *Apprenticeship in Critical Ethnographic Practice*. Chicago: University of Chicago Press.

Lave, Jean, and Etienne Wenger. 1991. *Situated Learning: Legitimate Peripheral Participation*. Cambridge: Cambridge University Press.

Mateos-Garcia, Juan, and W. Edward Steinmueller. 2008. "Open but How Much? Growth, Conflict and Institutional Evolution in Open Source Communities." In *Community, Economic Creativity, and Organization*, edited by Ash Amin and Joanne Roberts, 254–281. Oxford: Oxford University Press.

Reilly, Katherine M. A., and Matthew L. Smith. 2013. "The Emergence of Open Development in a Network Society." In *Open Development: Networked Innovations in International Development*, edited by Matthew L. Smith and Katherine M. A. Reilly, 15–50. Cambridge, MA: MIT Press; Ottawa: IDRC. https://idl-bnc-idrc.dspacedirect.org/bitstream/handle/10625/52348/IDL-52348.pdf?sequence=1&isAllowed=y.

Singh, Parminder J., and Anita Gurumurthy. 2013. "Establishing Public-ness in the Network: New Moorings for Development—a Critique of the Concepts of Openness and Open Development." In *Open Development: Networked Innovations in International Development,* edited by Matthew L. Smith and Katherine M. A. Reilly, 173–196. Cambridge, MA: MIT Press; Ottawa: IDRC. https://idl-bnc-idrc.dspacedirect.org/bitstream/handle/10625/52348/IDL-52348.pdf?sequence=1&isAllowed=y.

Smith, Matthew L., Laurent Elder, and Heloise Emdon. 2011. "Open Development: A New Theory for ICT4D." *Information Technologies & International Development* 7 (1): iii–ix. https://itidjournal.org/index.php/itid/article/viewFile/692/290.

Smith, Matthew L., and Ruhiya Seward. 2017. "Openness as Social Praxis." *First Monday* 22 (4). http://firstmonday.org/ojs/index.php/fm/article/view/7073.

Srinivasan, Janaki. 2011. "The Political Life of Information: 'Information' and the Practice of Governance in India." PhD diss., University of California, Berkeley.

Talja, S. 2010. "Jean Lave's Practice Theory." In *Critical Theory for Library and Information Science: Exploring the Social from across the Disciplines,* edited by G. J. Leckie, L. M. Given, and J. E. Buschmann, 205–220. Santa Barbara, CA: Libraries Unlimited.

Weber, Steven. 2004. *The Success of Open Source.* Boston: Harvard University Press.

Wenger, Etienne. 1998. *Communities of Practice: Learning, Meaning, and Identity.* Cambridge: Cambridge University Press.

Wenger, Etienne, Nancy White, and John D. Smith. 2009. *Digital Habitats: Stewarding Technology for Communities.* Portland, OR: CPsquare.

Wenger-Trayner, Etienne, and Beverly Wenger-Trayner. 2015. *Communities of Practice: A Brief Overview of the Concept and Its Uses.* http://wenger-trayner.com/introduction-to-communities-of-practice/.

Reflections I

5 Stewardship Regimes within Kenya's Open Data Initiative and Their Implications for Open Data for Development

Paul Mungai and Jean-Paul Van Belle

Introduction

Reilly and Alperin (chapter 2, this volume) argue that there are a variety of ways that open data can be connected to meaningful use, depending on the actors and stewardship regime that manage the data. The concept of stewardship adds to the open data scholarship by emphasizing open data intermediation and asking whether powerful actors engage in intermediation strategies that align with the types of social values that citizens prioritize. Thus, identifying stewardship regimes involves uncovering and confronting actors' power and position, values and relationships, and how and why the needs and wants of others (who might benefit from open data) go unmet.

In this reflection, we contemplate the potential of the stewardship approach to better understand an open government data initiative. We draw on research conducted on Kenya's Open Data Initiative (KODI). The initiative's purpose, as defined by the Kenyan government, focused on increasing access to government data sets by making them available in free and easily reusable formats, with the aim of increasing government accountability and transparency (ICT Authority 2017).

Kenya and Open Data

Kenya's government has made significant efforts toward becoming more open and participatory. This includes joining the Open Government Partnership (OGP) in 2011 and making commitments around three thematic areas: e-governance, legislation and regulation, and public participation (Open Government Partnership 2018).

On July 7, 2011, the president of Kenya, the Hon. Mwai Kibaki, established the Kenya Open Data Initiative (KODI) under the Kenya ICT Authority to help implement its OGP commitments. The initiative is managed by the ICT Authority, an agency that falls under the Ministry of Information, Communications and Technology. The agency provides technical support and training, helps create awareness of open data among the public and other government agencies, and manages the supply of open data from government and the demands for open data from the general public and government. The KODI portal (www.opendata.go.ke) contains open data search and visualization tools and an interactive feedback mechanism for data consumers.

The portal currently hosts more than nine hundred open data sets, including data on national or county expenditures, population, school enrollment, road maintenance reports, and water and sewage connections (ICT Authority 2020). Arguably, this demonstrates some success. However, there is still a need to improve the quality of open data, as this hinders usage and value. Moreover, shortfalls that had been observed in KODI include a mismatch between what is made available and what citizens want, outdated information, and the lack of readily usable data sets (Mutuku, Mahihu, and Sharif 2014). The experiences of government officials provide insight into the intermediation strategies that offered the greatest help in creating certain kinds of social value. To improve the social value of the initiative, KODI engaged in the intermediation strategies we discuss next.

Data Fellows as Key Open Data Stewards within KODI

Reilly and Alperin's stewardship approach suggests that we look at which actors are *stewarding* the data, which includes the production, distribution, and use of the data to create value. KODI developed new *data fellow* roles to facilitate data practices among both the government agencies who were contributing data to KODI and the citizens who wanted and needed KODI data. We investigated how and why these actors were stewarding the data and what kinds of value their actions were contributing.

KODI deployed data fellows to certain agencies for a period of six months to create more awareness of its activities, improve its relationship with contributing government agencies, and understand the data release cycle, which is necessary for developing the data release calendar and determining when to ask for updates. One data fellow, who was deployed at a county government office, spoke about his role of negotiating with the Finance

Department on behalf of the county office: "We spoke to the Chief Officer; he was the guy who was dealing with the raw finance data. We explained everything to him; he said we draft a general questionnaire of what we would like. After that he emailed us the data." The data fellow helped the county office staff define the requirements and communicate effectively with the Finance Department.

Data fellows were also responsible for helping departments responsible for releasing data understand their responsibilities and adopt standard data release procedures. For instance, the data release calendar outlines when government releases data throughout the year (ICT Authority 2020), which is meant to facilitate planning and routinization for contributing departments. However, even when departments deliver data sets according to this calendar, they still value support from data fellows to check for correctness. Once the departments send data sets to KODI, data fellows check for data correctness in consultation with the source of that data. For instance, one data fellow who was responsible for data acquisition at KODI described how verifying the correctness of data often ended in a final meeting with the data custodian who provided the data: "'You gave us this data set, we've done this with it, so what's your input?' They will say, 'No remove this, add this, [or] it's okay.'" This added layer of support and verification increases the quality of the data in the short term and helps data custodians learn, from the data fellows, what is needed at future data release intervals.

Nevertheless, KODI only publishes a data set if approval has been granted by the agency responsible for that data set. The open data release form helps formalize this approval. As one data fellow explained, "We also have the Open Data Release Form that they sign and a representative from our side also sign[s] and then we publish it … so generally this data remains the property of that particular agency." The form defines the owner of that data, who is referred to as the publishing partner (government ministry or parastatal), the primary contact person from the publishing partner, and KODI's reviewer of that data set. It also outlines the terms and conditions that KODI applies to the data set, including licenses, restrictions on how the data set may be altered, and the continued responsibilities of the actors in replenishing the data set. When data providers enter into a mutual agreement with KODI and its facilitators, such as the data fellows, they may derive a greater sense of the value of the activity in maintaining and complying with standardized procedures.

When linking the stewardship practices of the data fellows to the type of value they create, Reilly and Alperin's (chapter 2, this volume) intermediation models gave us an initial sense of what kinds of value might be generated by the fellows' actions. For instance, a large portion of a data fellow's role is to increase the quality of open data from a technical perspective. They achieve this using various techniques, including data release prioritization, data release forms, and by providing technical support and follow-up on data release as per the signed agreements with the agencies. They also help verify the correctness of the data with the data source. In this regard, they could be termed infomediaries, consistent with Reilly and Alperin's (chapter 2, this volume) arterial school, because of their active role in ensuring the flow of data. Yet, within much of the data fellows' explanations, they concentrate on assuring flows from producers without much attention to users. Our research therefore confirms that the data fellows' stewardship practices are focused on the raw and potential use value of data, without much consideration for how different kinds of value will manifest for users.

Nevertheless, some of the fellows, among other KODI officers, did focus on engaging citizens. The next section focuses on this link.

Actors at the Interface between KODI and Citizens

For many countries, including Kenya, government data are plentiful, scattered across multiple agencies, and, at times, duplicated. In the case of Kenya, for instance, there are eighty-three government agencies, each with some unique data sets. However, it takes considerable resources to publish data, and it is not clear that all data sets are equally useful. Thus, choices have to be made as to which data sets to prioritize.

KODI developed and adopted a range of strategies to engage with citizens to help with this prioritization. Some of the strategies, such as the Suggest Dataset feature, were not heavily facilitated, whereas engaging users through social media or by blogging involved more intensive facilitation by KODI actors, the results of which will be discussed later.

Regarding the use of social media to solicit citizen input, actors (including data fellows and KODI officers) focused on seeking suggestions regarding what data to produce. They also promoted public events that may be of benefit to citizens. For instance, one of the data fellows appointed by the ICT Authority to provide technical support at the National Transport and Safety Authority (NTSA) spoke about how he used social media to publicize an event: "What I would do if we have any engagements like [a] workshop, I would post that on Facebook and Twitter, then you find from there people request; they say, 'It's good you have this, how come you don't have that?' So, from there you're able to know. Then also guys at the host institution, there's a way they say, 'I think this is what the people need to know.' So, probably that also helped." By publicizing the event, the fellow opened a channel of communication that was used to gather insight into what citizens wanted. However, this suggests that the fellow also needed to be able to act on this information. When the fellow speaks about the host institution and their interpretation of what citizens will need, the fellow operates as both an advocate and bridge between the two perspectives.

It is more common for the team at KODI to help citizens find data from the Kenya Open Data Portal. This support is necessary since most of the users do not have the skills to effectively retrieve a data set or generate the desired information using the tools available on the portal. When speaking with an interested citizen, a KODI data analysis officer remembered, "He was saying, 'Money Safaricom pays to Airtel [LAUGH].' I am like, 'Have you tried looking for Termination Rates or something like that?' ... When he looks for Termination Rates he discovers it." The officer gave another example of a citizen he had spoken with that week to find a data set. They both needed help to reframe the questions they were interested in answering to align with the data and terminology used by KODI. Had the officer not been familiar with the data sets available or capable of interpreting their requests, the portal would have had very little value for these citizens.

The data analysis officer also highlighted a blogging strategy they were using to further assist citizens in using the portal's visualization tools. Usually, having the portal designed with visualization tools helps to make data more accessible to a broader audience. However, data sets are often difficult to make sense of, even with the use of visualization tools. The officer explained how blogging insights might shed some light on how citizens can use the visualization tools effectively:

> What we are doing this year is something we are calling the significant number, so significant number is where we pick a data set, do a blog post about it, but then try and point [out] some numbers that we find very interesting. For example, this week's significant number is the, what do you call it? [road carnage] The most ... road accidents in Kenya occur at 6 PM, which is very insane because I am thinking everybody is stuck in traffic at 6 PM. Actually, not accidents but fatality,

> like people die more at 6 PM. Looking at Nairobi there's too much traffic and Nairobi contributes the bulk of road accidents in Kenya, doing something about 50%–60%. So, when we consider that, why are all these people dying and traffic is very slow at 6 PM? So, it's a very interesting fact, so that's entirely what significant number is all about.

The blog posts give examples of how data sets, and visualizations of them, can be used to tell compelling stories and thus promote data-driven journalism and advocacy. This strategy may help citizens value the data sets and visualization tools differently, so that they start to explore the portal as well.

Lastly, KODI, in partnership with civil society actors, developed laws and policies to ensure the sustainability of open data access. Prior to this, the media had been denied access to some contentious government information on the grounds of confidentiality or secrecy (The Standard 2016). The largest milestone in the effort to create supportive laws and policies was the passage into law of the Access to Information Act in 2016 (Access to Information Act 2016). KODI and civil society actors, such as the International Commission of Jurists-Kenya, Transparency International, and Article 19, collaborated in drafting and lobbying for laws on access to information and data protection. As the KODI project coordinator stated: "I am now heavily involved in writing policy for the project. So we did the Access to Information and Data Protection Bill that are both in Parliament now. On that we are now writing an open data policy, which is to guide how people should make data available. We are talking about machine readable sort of data. So, I have been involved in writing policy as well and meeting high level management people to create that."

The Data Protection Bill seeks to provide protection of personal information. It was drafted in 2012 and was debated in the Parliament of Kenya for the first time in 2013. More than six years later, it was passed into law on November 25, 2019 (Republic of Kenya 2019). In addition to these laws, KODI is currently developing the first national Open Data Policy in consultation with the various stakeholders, including other government agencies, the private sector, and civil society.

In summary, some of the KODI data fellows' actions could belong to the bridging school because their role was to mediate between the sources of data and the end users through the portal. The data analysis officer also helped translate data through the significant number blog, which highlighted some of the data sets. Data fellows also conduct public training and workshops that are aimed at creating awareness and building capacity on how to search and translate data into meaningful information. These actors therefore help to bridge scientific data and bureaucratic complexities that users would otherwise face while trying to access some of these data sets with meaningful use cases for members of the public.

Likewise, KODI actors were also involved in drafting and advocating for the enactment of open data laws and policies and coordinating the implementation of these policies within government. In addition, KODI actors also helped to forge relationships between government and other stakeholders, including civil society, the developer community, and the private sector, which includes media companies such as Nation Media Group.

Connecting KODI's Open Data Stewardship Regimes to Open Data for Development

Reilly and Alperin's (chapter 2, this volume) stewardship approach offers a new discourse for open data for development. For instance, the intermediation models presented provide a new set of terms that can be used to understand the choices actors make, where their contributions lay, and how and why these create different kinds of social value. We examined data fellows and KODI staff as key actor groups and interpreted their experiences to help us understand what was going on inside the initiative.

Data fellows engaged in a number of intermediation strategies, sometimes simultaneously, including those that could fall into the arterial school, the bridging school, and the ecosystems school, whereas other KODI staff, such as the data analysis officer, may have had more targeted roles focused on specific intermediation strategies, such as the significant number blog. Most of these intermediation strategies emphasized the institutionalization of open data practices within the Kenyan government, which indicates a commitment to the wider social benefit of open data. In contrast, other intermediation strategies targeted communicating with citizens and enabling their deeper engagement with open data. However, the value created through these strategies also seemed to focus on the wider social benefit of open data rather than on the specific use value for different groups of citizens. This may facilitate buy-in, but not uptake, of open data in meaningful ways.

As a discourse, it is important for researchers to push past surface-level understanding of stewardship regimes and the types of value they create within open data initiatives. In other words, we might now question how we can use this knowledge to build a better understanding of the factors that link KODI to positive social transformation. Reilly and Alperin (chapter 2, this volume) remind us that if we view open data for development as something that actors make and do, and thus view it as shaped by the values actors wish to extract from data resources, we have no choice but to focus on meaningful use, which requires a greater understanding of the actors largely absent from our analysis. According to Samoei et al. (2015), who calculated poverty rates based on 2005–2006 household surveys, almost half (45.2 percent) the population was living in poverty. For the average citizen, navigating intersecting structures of power, such as class, gender, and education, makes their experience with open data for development more challenging. The stewardship approach urges us to shine a flashlight on those processes that may further entrench asymmetrical or inequitable public engagement with open data.

While other approaches may instead focus on the agency of open data users and on understanding their needs and wants, the stewardship approach seeks to interrogate and highlight the mechanisms of institutionalization of open data distribution processes. Learning about the KODI actors' stewardship regimes and the impact of these regimes on the institutionalization of KODI within the Kenyan government is a key way to understand whether the main stewards of open data take into consideration the average citizen's values and, if so, how.

References

Access to Information Act. 2016. "Access to Information Act, no. 31 of 2016." https://www.cuk.ac.ke/wp-content/uploads/2018/04/Access-to-Information-ActNo31.pdf.

ICT Authority. 2020. "Open Data." http://icta.go.ke/open-data/.

Mutuku, Leonida, Christine M. Mahihu, and Raed Sharif. 2014. "Exploratory Study on the Role and Impact of Kenyan Open Data Technology Intermediaries." Open Data Research Network briefing note, July 10. http://www.opendataresearch.org/content/2014/671/briefing-note-exploratory-study-role-and-impact-kenyan-open-data-technology.html.

Open Government Partnership. 2018. Government of Kenya Open Government Partnership National Action Plan III 2018–2020. Washington, DC: OGP. https://www.opengovpartnership.org/wp-content/uploads/2018/12/KENYA_Action-Plan_2018-2020_0.pdf.

Republic of Kenya. 2019. Kenya Gazette Supplement. The Data Protection Act, 2019. Nairobi: Republic of Kenya. http://kenyalaw.org/kl/fileadmin/pdfdownloads/Acts/2019/TheDataProtectionAct__No24of2019.pdf.

Samoei, Paul K., Samuel Kipruto, Mary Wanyonyi, David Muthami, and John Bore. 2015. *Spatial Dimensions of Well-being in Kenya*. Nairobi: Kenya National Bureau of Statistics.

The Standard. 2016. "Law on Access to Information Was Long Overdue." *Standard Digital*, September 3. https://www.standardmedia.co.ke/article/2000214485/law-on-access-to-information-was-long-overdue.

6 Changing Infrastructure in Urban India: Critical Reflections on Openness and Trust in the Governance of Public Services

David Sadoway and Satyarupa Shekhar

Introduction

Cities in India are in a state of flux characterized by rapid changes in population, land use, and infrastructural arrangements. With approximately 68 percent of its nearly 1.21 billion residents still living in rural communities (Census of India 2011a), the relatively recent rapid growth in India's cities has exerted severe pressure on local governments to better supply public services.[1] Indian cities can be understood as vast *provisioning machines* (Amin 2014) that provide services and infrastructure for sustaining the lives of their citizens (figure 6.1). In this critical reflection, we discuss how questions about *open systems* and *trust*—elaborated on in the theoretical work of Rao et al. (chapter 3, this volume)—relate to the provision of urban services and infrastructure. Internationally, a variety of open practices and systems demonstrate apparent promise for improving urban public service delivery. For example, governments and civil society groups have created open platforms and have crowdsourced citizens' input on diverse issues linked to local service or infrastructure needs (Hagen 2011).

Our research—drawing on perspectives of both local government and civil society intermediaries—provides insight into public service and infrastructure issues in a rapidly changing city in India, as well as theoretical reflections for advocates and theorists of open systems. We link our study to a critique of Rao et al.'s operating theory, discussed in chapter 3 of this volume, about *trust* (or trustworthiness) in combination with *open systems* (or openness), and we apply this to questions about the provision of public services and infrastructure in Chennai, India. Rao et al. (chapter 3, this volume) have introduced a trust model that applies to open systems in a

Figure 6.1
Leaky water pipe in south Bengaluru (Bangalore).
Source: Sadoway, Gopakumar, and Sridharan (2013).

generic sense but also, they suggest, can be applied to service provisioning. Indeed, the study of trust in the development of cities and urbanization has important relevance, as Tilly's (2010) historical work on the development of urban *trust networks* suggests. His work identifies how the earliest cities were both shaping and shaped by struggles over their residents' mutual trust commitments. This leads to the question of what *trust* and *openness* actually refer to in relation to the provision of urban services. Chopra and Wallace (2003, 2) conceptually suggest that questions about *trust* involve three interrelated elements: "*a trustee* to whom the trust is directed, *confidence* that the trust will be upheld, and a *willingness* to act on that confidence" (italics original). On the other hand, *open praxis*, according to Smith and Seward (2017), involves both processes and practices of knowledge governance that are free and nondiscriminatory, or open to participation.[2]

Our research, however, conducted with a variety of intermediaries in Chennai, makes us skeptical about whether current forms of digitally inspired open development—especially approaches led or seeded by external sponsors—are being devised in ways that address key local servicing needs. We raise questions about Rao et al.'s (chapter 3, this volume) trust model because it positions publics as disembodied feedback channels (that is, as *external* agents in *sponsored* systems and/or *information generators*) rather than as (pro)active citizens or comanagers of information. Importantly, their model arguably downplays the complexities of service provisioning, particularly where aspects of overlapping or multilevel governance remain the norm (that is, various government bodies and agencies as well as civil society groups involved in questions about public services). While Rao et al. (chapter 3, this volume) refer to "trust in the sponsor," our research highlights the polycentric, multilevel power dynamics that shape complex local governance arrangements (not just single-level sponsorship). Furthermore, our findings highlight the politics of outsourcing or offloading of public service sponsorship (and trust) or management to private or nongovernmental organizations (NGOs), including recent debates in India about the provision of either free or *sharing economy* services.[3] While Rao et al.'s (chapter 3, this volume) trust model identifies broad power dynamics, we suggest that questions about specific *power trade-offs*—such as understanding why local infrastructural and servicing power struggles are occurring and how public collective or universal services are being undermined by private provisioning proposals—remain crucial to understanding open and trustworthy modes of infrastructure and services governance.

Our investigation ultimately focused on perceptions of trust and the importance of openness in the provision of public services—such as bus shelters, public libraries, water, streetlights, and so forth. To do this, we employed three overarching questions to investigate the nature and context of service provision in Chennai: How are public services and infrastructural provisions being governed? Can open practices improve the governance of urban public services and infrastructure, and how? And how are trust relations affecting current service provision practices? We conducted semistructured interviews in 2016 and 2017 with twenty-four

Chennai-based government officials, staff, elected councilors, and civic association intermediaries.

The remainder of this reflection explores our findings on public service provisioning in Chennai and ends with our critique of Rao et al.'s (chapter 3, this volume) trust model.

Chennai as a Provisioning Machine

Chennai has a metropolitan population of 8.69 million residents.[4] It is also an iconic economic gateway to the state of Tamil Nadu and southern India (Sood 2013, 95) and has been dubbed "the Detroit of India" for its growing strength in vehicle manufacturing (Krishnamurthy and Desouza 2015, 118) (see figure 6.2). The rapid rise in population, automobile use, and land use changes have all put heavy pressure on Chennai's public services and infrastructure. The Corporation of Chennai (CoC), which Sridhar and Kashyap (2012, 99) identify as the "oldest corporation in India," founded in 1688, is the civic body that governs Chennai. The CoC government is led by a mayor and a group of councilors elected from two hundred electoral wards across the city. However, like other large cities in India, the Government of India (GOI or Centre) and the state government play a dominant role in local urban infrastructural governance and in steering the provision of services.

Public Service in a Multilevel Governance Reality

> A public service is a service where citizens should consider themselves as partners of the service. Citizens right now see themselves as consumers and not as participants.
>
> —Respondent 3, member of a civil society organization, interviewed on December 22, 2016

The role of cities in relation to the states and Centre is symptomatic of the longstanding problem of aborted decentralization in India. Observers have linked the longstanding lack of decentralization of financing, professional staffing, and planning in Indian urban governance (Mukhopadhyay 2006; Sivaramakrishnan 2007; Sivaramakrishnan 2011) with the corollary of increasingly concentrated power in New Delhi and state capitals.[5] Such maldistribution of political power remains a crucial impediment to

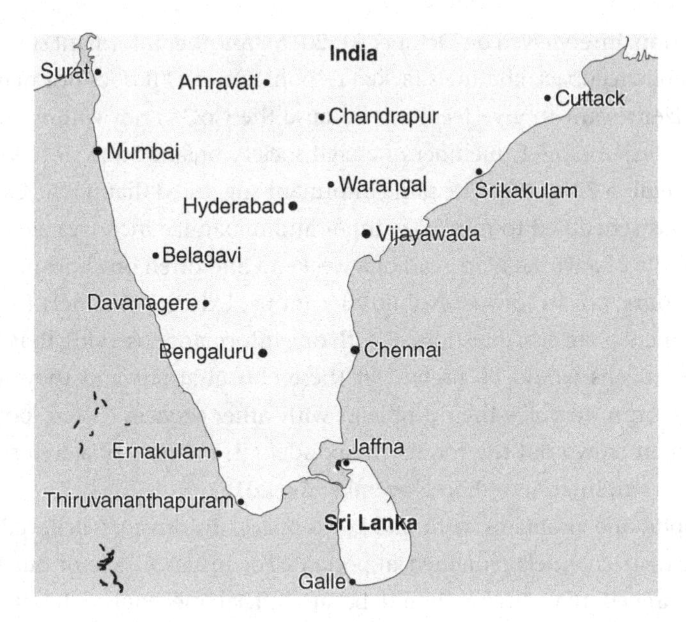

Figure 6.2
Map of southern India showing Chennai (Madras) on the southeast coast.
Source: Open Street Map (2018), https://www.openstreetmap.org/#map=5/12.983/73.960&layers=T.

building trust and potentially initiating open practices between (and for) citizens and local governments. Related to this, Krishnaswamy, Idiculla, and Champaka (2017, n.p.) argued that "power should be located as close to the people as possible in the smallest political units feasible." This suggests that *subsidiarity*, or the act or practice of decentralization in a governance system, potentially enables an "alignment between democratic authority and urban planning power" (Krishnaswamy, Idiculla, and Champaka 2017, n.p.). Indexing the degree of subsidiarity in governance—particularly legal, political, fiscal, or economic—can thus provide insights into the ability (or autonomy) of local governments to shape urban public service provision.

Ideally, open digital practices—such as introducing public feedback channels—would contribute to better aligning democratic powers with public service provision; however, our respondents expressed some skepticism about this. As one informant stated, "e-services can help ease the process of getting things done, but they cater only to the educated and middle- and upper middle-class people" (Respondent 9, resident welfare association, interviewed on October 11, 2016). Another informant suggested that public feedback channels lacked responsiveness: "It does not matter if the citizens want to give feedback because the CoC is not willing to take them" (Respondant 1, member of a civil society organization, interviewed on November 22, 2016). The same informant suggested that public engagement was scheduled to minimize input and maximize inconvenience ("in the middle of a workday instead of a weekend and often on short notice of a few hours") or in low-accessibility locations. Existing channels for civic engagement were also questioned, with one informant observing that "only retired citizens would be present at these consultations and they would use the forum to voice their problems with other services. Often, political henchmen crowd out the room" (Respondent 1, member of a civil society organization, interviewed on November 24, 2016).

Despite the problems with public feedback, informants believed that engagement channels remained important. For instance, one of our informants argued that "there should be an official mechanism to organize residents of various neighborhoods to discuss civic issues and make representations to their elected representatives" (Respondent 6, member of civil society organization, interviewed on November 10, 2016). Another informant noted that "there are mechanisms like the mayor's meeting every Monday morning or the online complaints cell, but these do not work. The city needs more decentralized mechanisms for a feedback system to work" (Respondent 3, member of civil society organization, interviewed on December 22, 2016).

Citizen Action Group (CAG), the Chennai civic association we partnered with for this research, also identified an overdependence on centrally appointed Indian Administrative Service (IAS) staff in providing public services. The IAS staff members serve on a rotating basis in local government offices. Rather than developing Chennai-based capacity in the CoC to address local needs, rotating staff or consultancies are responsible for public services and infrastructure. Even within the CoC, subsidiarity, such as greater local ward feedback mechanisms or powers, is severely lacking. This was exemplified in the statement made at the beginning of this section by informant R3, who called for decentralization to neighborhoods. Models such as Rao et al.'s (chapter 3, this volume) should therefore account for the multilevel power struggles that influence setting of public priorities for urban services.

Valorizing Corporate and Consultant-Driven Service Solutions

A main finding of our research revolved around India's design and implementation of its Smart Cities Mission (SCM). SCM is a top-down initiative directed by the central government of India, and its formulation and financing favors corporate and consultant-driven solutions. SCM is also focused on middle-class concerns, such as parking, rather than basic needs, such as water provisioning. The high valorization of *smart cities* and high-tech solutions arguably is linked to a fetishization of build-operate-transfer and public-private partnership models in Indian cities. Such approaches defer to external expertise for how urban public services and infrastructure nominally ought to operate (Coelho, Kamath, and Vijaybaskar 2011; Sadoway et al. 2018; Sangita and Dash 2005). The valorization of corporate and consultant-driven solutions also serves to diminish trust in the longstanding local knowledge systems and local staff capacity.

The Chennai Smart City (CSC) initiative is emblematic of public-private partnership models, as it favors corporate, technology-oriented solutions over democratically governed service provision. The CSC initiative's proposal was prepared by the global consulting firm Jones Lang Lasalle Inc. One claim in this proposal was that extensive public consultations—including with elected representatives and NGOs—were conducted. However, the proposal indicates that only the opinions of the CoC mayor, a single member of the Legislative Assembly, and just two business-oriented civil society organizations—the Institute for Transportation and Development Policy (ITDP Chennai) and Chennai City Connect—were involved. There was, concomitantly, limited public engagement. Moreover, software and technology vendors were consulted, and their suggestions focused on technology-oriented solutions employing sensors, chips, or cameras, while largely ignoring local basic service and more basic infrastructure needs, such as water, sewerage, and mass transit. One informant, for instance, suggested that "right now, there is a perception of what people want and ideas like the elevated expressways, or RFID [radio frequency identification] tagging garbage bins, are proposed" (Respondent 3, member of a civil society organization, interviewed on December 22, 2016).

While information technologies could clearly be employed in augmenting or potentially improving provision of any public service, the concerns of our informants centered on the belief that these technology- and consultant-led approaches were premature to the extent of missing the need for basic service and infrastructure needs across the city. Another respondent raised concerns about the improper distribution of basic services and how these services were being provided to neighborhoods on an ability-to-pay basis rather than being universally affordable for residents, saying, "There are some people who can afford to pay, but there are others who are not able to afford [to], yet officials demand that they pay for all services" (Respondent 9, resident welfare association, interviewed on November 10, 2016). Kundu (2011), from a public investment perspective, has traced how infrastructure investments in India favor affluent neighborhoods and the cities most able to (re)finance cost recovery. These comments and observations highlight the chasm that needs to be overcome when designing open systems that put local priorities for public services or infrastructure first, let alone devising trustworthy technologically supported solutions.

Some CoC staffers directed civil society groups to consultants when they sought information. Our observations indicate that CoC staff are transferring their responsibilities to consultants and are losing their institutional capacity to service local communities in the process. For example, one informant noted that "the engineer managing the project would also not know or be able to recall what the figures are. Hence every time we required any data, the engineer would connect them to the concerned consultant" (Respondent 1, member of a civil society organization, interviewed on November 24, 2016). One official explained that CoC engineering staff "have support from consultants, since these days a lot of projects see the involvement of external parties. Though it is an opportunity for officials and staff to pick up new skills, they leave it to the consultants to do the job" (Respondent 21, city government official, interviewed on February 24, 2017). This also relates to our earlier observations about subsidiarity, in part because central and state-level programs can valorize the professionalism of external or private sector consultants rather than developing in-house or homegrown public service talent.

Another consequence of valorization (of consultants and corporations) is that the CoC appears to be treating citizens deferentially as passive and disconnected consumers rather than as engaged political participants. One informant observed that "there is a lot of disconnect between the government and the citizens, [and] with extremely high use of ICT-based infrastructure [(information and communication technology)], completely useless and unnecessary ideas are approved and executed, resulting in a major waste of public money" (Respondent 3, member of a civil society organization, interviewed on December 22, 2016). This raises questions about how programming for public services is being devised, funded, and approved—and whether governments favor corporate and consultant-driven solutions over more universal, democratic, and collaborative service provision. As external governmental infrastructure financing projects and external consultant-driven approaches become further entrenched in cities like Chennai, open practices would appear to be more difficult to devise.

Work-Arounds for Opening Up Public Service Accessibility

Based on the preceding discussion, we suggest that open system advocates and theorists need to consider how their approaches could not only open up or increase the accessibility to and setting of priorities for public services but also how their approaches might unwittingly limit or misdirect access to such services. Despite the major challenges hindering democratic and collaborative service provision discussed earlier, some citizens and public officials are finding pathways for accessing services. At times, these improvisations involve developing ad hoc solutions or adaptive or situational workarounds. Such workarounds have implications for how public service provision functions and how service provision systems may shift over time. However, we do not wish to romanticize civic or local government workarounds as necessarily innovative service provision models. Instead, we highlight them as features of public service systems that signal a lack of empowerment and trust-building among local citizens. While we have limited space for elaboration here, examples of trust-building from our interviews included comanagement of problems with government and residents during the 2016 floods, the use of direct public dial phone connections to CoC officials for improving access and accountability, and civil society groups working both with and also independent from government to address information, infrastructure, or service asymmetries. For example, during the 2016 flooding disasters that beset Chennai, one respondent suggested that there was a mutual appreciation of local residents' needs by CoC officials, as noted in the following observation (Respondent 9, resident welfare association, interviewed on November 10, 2016): "They worked with us like common people without thinking that they were government officials. But now the very same people show their authority and attitude when I approach them for any work."

After the disaster and the common or partnership mode of governance, the informant speaking about the flooding suggested that there was a return to paternalistic approaches. Issues of fairness in service provisioning were also reflected by another informant's comments: "There is also no equity in CoC. Current and retired public officials have clout so their complaints are attended to immediately, even from senior engineers" (Respondent 24, retired city government official, interviewed on February 16, 2017). This highlights how service asymmetries can be shaped by local personal networks that also undermine the possibility of building or strengthening trust networks among wider publics.

Since mobile phones have become an omnipresent part of familial or social networks, the use of direct public dial phone connections to CoC officials for improving service access and accountability suggests another workaround that has opened up the situation for some residents. In other situations, where some communities have been unable to access services, wealthier or more connected communities—such as those with active resident welfare associations (RWAs)—have also devised workarounds to address their needs. Workarounds for those with powerful political or staff connections were illustrated in these comments (Respondent 2, member of a civil society organization, interviewed on January 10, 2017): "We do not find the need to interact with elected representatives like councilors or ministers. We have some eminent residents of the city who are part of our [RWA] Board and who accompany us to meetings with senior officials."

Our colleagues at CAG also observed that in Chennai several affluent neighborhoods demanded garbage collection twice a day, while many low-income areas have this service only once a week or every two weeks. It was also observed by our colleagues that repairing roads, water supplies, and electricity faults in Chennai has also been shaped by the influence exercised by wealthier communities. The workarounds that these RWAs have devised could hardly be described as adequate solutions for accessing what an IAS official in Chennai (Respondent 21, city government official, interviewed on February 24, 2017) described as "rights-based services," since many others remain unable to exercise their right to access public infrastructure.

Issues of infrastructure access revolve around public service provision, since some citizens or groups appear to have access and others simply do not. Despite the notable power imbalances, our findings also suggest that there remains some hope for more community collaboration, power sharing, or governance innovation through workarounds. This was perhaps evident in one type of civic workaround that a local group used to generate their own data as an alternative to inadequate public information, as noted in the following comment (Respondent 4, member of a civil society organization, interviewed on January 5, 2017): "For the various studies we conducted, the CoC did not have the level of data we required.... We had to first create a Detailed Project Report which included a technical, financial, social, and environmental study so that we could gather data." This anecdote suggests that some groups were opting for workarounds to garner data for achieving improved public services or public responses. Workarounds were also identified in the local tendency for quick fixes among public agencies, as noted in the following observation: "All these agencies do '*jugaad*' [meaning a quick fix] that usually does not completely solve the problem" (Respondent 1, member of a civil society organization, interviewed on December 22, 2016).

Workarounds, as implied in our earlier observations on valorization and subsidiarity, suggest that distinct local sociocultural and political histories shape Chennai's service and infrastructure conditions. In the following section, we relate the Chennai case to the questions about trust and openness that we raised at the beginning of this discussion.

Reflections on Public Services and Open, Trustworthy Systems

In this short reflection, we have argued that Rao et al.'s (chapter 3, this volume) trust model requires a deeper focus on local contextual complexities related to public services and infrastructure provision and priorities. Our research in Chennai suggests that research into trust and openness needs to deeply consider the *local power struggles* over urban service needs and provisioning as well as *priority setting*, especially where there are diverse and changing local priorities. The observations made by our civil society intermediaries and officials within the CoC affirmed this to some extent; however, additional research and more varied perspectives would be helpful.

On the surface, there also appears to be potential for opening up channels for citizen input (such as crowdsourcing) to address concerns about local service deficiencies. However, as our previous work in Chennai has found (Sadoway and Shekhar 2014), transformations in urban infrastructure governance are unlikely to occur unless basic needs and community-defined priorities are first addressed, particularly via electoral political mechanisms, as well as being embedded in local knowledge systems. Our research has also highlighted a need to analyze the overlapping or multilevel power dynamics—among governments, civil society, and business—not just trust in single-level sponsorship, which is implicit in Rao et al.'s (chapter 3, this volume) approach.

Additionally, our findings underscored the importance of understanding how advocates for deepening the role of open systems and improving trust in the governance of urban services need to consider questions about the degree of local subsidiarity, the nature of corporate or consultant-driven solutions in a given context, and the types of local workarounds that alter or reshape urban service provisioning or provisions. Overall, we found the Rao et al. (chapter 3, this volume) model underequipped for analyzing the complexities of urban service provisioning, particularly in fast-changing Indian cities and in city regions where multilevel or polycentric governance remains the norm. For example, their view of the public as disembodied feedback channels (*external* agents in *sponsored* systems and/or *information generators*) does not capture the dynamics of the public as (pro)active citizens or comanagers of information, as our short study in Chennai identified. This also highlights the dangers of outsourcing or offloading service responsibilities not only to consultants but also to private, charitable, or nongovernmental vehicles. We posit that greater citizen control (or cocreation) ought to play an integral role in analyzing or actualizing open governance practices. In the view of many of our informants—particularly civil society intermediaries who work on the front lines with diverse communities—Chennai's materially poorest residents (and also many in the growing middle class) appear to be largely left out of local civic engagement regarding future land use or infrastructure planning and budgeting.

Instead of simply focusing on improving public services, we have suggested that the first step in improving services and *building urban trust networks* would be to focus on directly involving the public first to address basic local public service and infrastructure needs and priorities, such as public and community toilets, water supplies, parks, child care facilities, or primary health care centers. One of our informants referred to these as "rights-based services" that needed to be "requested by communities, rather than individuals" (Respondent 21, city government official, interviewed on February 24, 2017). This illustrates how challenges about public service provision have resulted in continued calls for "shared infrastructural rights" (Amin 2013, 486) alongside new forms of democratic practices in India's dynamic cities (Coelho, Kamath, and Vijaybaskar 2011, 30). However, as Tilly's work (2010, 272) on trust highlights, while cities can serve as platforms for competing trust networks, their ability to properly integrate democratic urban governance with the provision of public services remains historically rare.

Finally, we suggest that further research is needed to explore how both *rapid urbanization* and *new urban citizenships* are reshaping not only service or infrastructure expectations but also questions about trustworthiness and openness in local governance. The politics of urban infrastructure and services therefore needs to be understood in relation to how both local servicing asymmetries and the local political contexts of class, caste, and gender intersect to (re)shape urban government.

Notes

1. We define *public services* as nominally universal services, governed or managed by public bodies, and provided to residents or citizens through a range of infrastructure. We define *infrastructure* as sociotechnical "assemblages of public works, technical installations, and institutional arrangements that mediate flows of services," such as water, waste, energy, mobility, and communications (Sadoway et al. 2013, 3).

2. Smith and Seward (2017) list four key practices: peer production, crowdsourcing, sharing, and consumption (for example, reuse, remixing, or repurposing).

3. An example of a private-led sharing economy service and infrastructure initiative is Facebook's Free Basics initiative in India. The initiative involves proposals by Facebook to bundle free online services, on an open platform, with free Internet access (Yim, Gomez, and Carter 2017).

4. Census of India (2011b) data.

5. This included governments ignoring the provisions of the Indian Constitution's 74th Amendment, which mandated the decentralization of functions to local governments and community wards' committees (Kundu 2011).

References

Amin, Ash. 2013. "Telescopic Urbanism and the Poor." *City* 17 (4): 476–492.

Amin, Ash. 2014. "Lively Infrastructure." *Theory, Culture & Society* 31 (7–8): 137–161. http://www.stellenboschheritage.co.za/wp-content/uploads/Theory-Culture-Society-2014-Amin-137-61.pdf.

Census of India. 2011a. *Primary Census Abstract: Figures at a Glance, India*. http://www.censusindia.gov.in/2011census/PCA/PCA_Highlights/pca_highlights_file/India/5Figures_at_glance.pdf.

Census of India. 2011b. *Provisional Population Totals, Urban Agglomerations/Cities Having Population 1 Million and Above*. http://www.censusindia.gov.in/2011-prov-results/paper2/data_files/india2/Million_Plus_UAs_Cities_2011.pdf.

Chopra, Kari, and William A. Wallace. 2003. "Trust in Electronic Environments." In *Proceedings of the 36th Hawaii International Conference on Systems Science*, Big Island, HI, January 6–9, 2003, 1–9. Washington, DC: IEEE Computer Society.

Coelho, Karen, Lilitha Kamath, and M. Vijaybaskar. 2011. "Infrastructures of Consent: Interrogating Citizen Participation Mandates in Indian Urban Governance." IDS Working Paper Series 362. Brighton: Institute of Development Studies.

Hagen, Erica. 2011. "Mapping Change: Community Information Empowerment in Kiberia (Innovations Case Narrative: Map Kiberia)." *Innovations* 6 (1): 69–94.

Krishnamurthy, Rashmi, and Kevin C. Desouza. 2015. "City Profile: Chennai, India." *Cities* 42:118–129.

Krishnaswamy, S., M. Idiculla, and R. Champaka. 2017. "The Scales of Planning: Adopting a Multi-scalar Approach." *The Hindu*, March 19.

Kundu, Debolina. 2011. "Elite Capture in Participatory Urban Governance." *Economic & Political Weekly* 46 (10): 23–25. http://re.indiaenvironmentportal.org.in/files/urban%20governance.pdf.

Mukhopadhyay, Partha. 2006. "Whither Urban Renewal?" *Economic & Political Weekly* 41 (10): 879–884.

Open Street Map. 2018. Openstreetmap.org. https://www.openstreetmap.org/#map=5/12.983/73.960&layers=T.

Sadoway, D., G. Gopakumar, V. Baindur, and M. G. Badami. 2018. "JNNURM as a Window on Urban Governance in India: Its Institutional Footprint, Antecedents and Legacy." *Economic & Political Weekly* 53 (2): 71–81. https://smartnet.niua.org/sites/default/files/resources/sa_liii_2_130118_rua_david_sadoway.pdf.

Sadoway, D., G. Gopakumar, and N. Sridharan. 2013. "Critical Urban Infrastructure(s): Launching and International Research Network on Critical Issues in and about Urban Infrastructure." Conference paper for "Engaging Canada: Emerging Priorities for Sustainable Partnerships," Shastri Indo-Canadian Institute, New Delhi, June 1, 2013.

Sadoway, David, Govind Gopakumar, Vinay Baindur, and Madhav G. Badami. 2013. "Assembling Infrastructure Decongestion: An Overview of Critical Issues in and about Urban Infrastructure and JNNURM in India." Seminar at National Institute of Urban Affairs, New Delhi, August 1, 2013.

Sadoway, David, and Satyarupa Shekhar. 2014. "(Re)prioritizing Citizens in Smart Cities Governance: Examples of *Smart Citizenship* from Urban India." *Journal of Community Informatics* 10 (3). http://ci-journal.net/index.php/ciej/article/view/1179/1115.

Sangita, Satyanarayana N., and Bikash C. Dash. 2005. "Electronic Governance and Service Delivery in India: Theory and Practice." Working Paper 165. Nagarabhavi, Bengaluru (Bangalore): Institute for Social and Economic Change.

Sivaramakrishnan, K. C. 2007. "Municipal and Metropolitan Governance: Are They Relevant to the Urban Poor?" In *The Inclusive City: Infrastructure and Public Services for the Urban Poor in Asia*, edited by Aprodicio A. Laquian, Vinod Tewari, and Lisa M. Hanley, 278–302. Washington, DC: Woodrow Wilson Center Press.

Sivaramakrishnan, K. C. 2011. *Re-visioning Indian Cities: The Urban Renewal Mission.* New Delhi: SAGE Publications.

Smith, Matthew L., and Ruhiya Seward. 2017. "Openness as Social Praxis." *First Monday* 22 (4). https://firstmonday.org/ojs/index.php/fm/article/view/7073.

Sood, Ashima. 2013. "Urban Multiplicities: Governing India's Megacities." *Economic & Political Weekly* 48 (13): 95–101.

Sridhar, K. S., and N. Kashyap. 2012. *State of India's Cities: An Assessment of Urban Conditions in Four Mega Cities.* Bangalore: Public Affairs Centre.

Tilly, Charles. 2010. "Cities, States, and Trust Networks: Chapter 1 of *Cities and States in World History*." *Theory and Society* 39 (3–4): 265–280.

Yim, Moonjung, Ricardo Gomez, and Michelle Carter. 2017. "Facebook's 'Free Basics' and Implications for Development: IT Identity and Social Capital." *Journal of Community Informatics* 12 (2): 217–225. http://www.ci-journal.net/index.php/ciej/article/view/1321.

7 Learning through Participation in a Weather Information System in West Bengal, India

Linus Kendall and Purnabha Dasgupta

Introduction

In India, rural livelihoods are crucially important, as they employ over 70 percent of the population; however, increasingly they are being adversely impacted by climate change (World Bank 2014). Agriculture in semiarid regions is especially affected, as it is dependent on seasonal rainfall. Pumped irrigation is both costly and difficult because of water scarcity as well as the dry climate. Small-scale and marginal farmers are particularly vulnerable, as they have limited economic, educational, or informational resources to allow them to adapt to climate change.

In 2015, the Development Research Communication and Services Centre (DRCSC), a nongovernmental organization based in Kolkata, West Bengal, India, began implementing a weather information system to help farmers adapt to climate change. The project is funded by the Adaptation Fund[1] via the Indian government's National Bank of Agriculture and Rural Development (NABARD) (Adaptation Fund Board Secretariat 2015). The weather information system project is designed to (1) provide weather forecasting and crop advisory services to nine thousand farmers in two blocks[2] of two districts and (2) support farmers in making use of these services in their agricultural practices. A complementary objective of the project is to develop a model for accurate microscale weather predictions within the targeted areas. Providing forecasts and advisory services, as well as developing a long-term weather model for the targeted districts, is intended to support farmers in adapting their livelihoods and agricultural practices to changing weather patterns. Additionally, the project promotes sustainable agricultural practices by encouraging limited use of inputs such as synthetic pesticides and fertilizers. Developing sustainable farming practices is important, too, as they mitigate the negative impacts of Green Revolution farming practices, which have become increasingly evident (Pingali 2012).

The project has placed farmers as active participants in the consumption, production, and distribution processes taking place within the weather information system. The assumption is that the more they actively contribute, the greater the chance that they will adapt their farming practices as a result of this engagement. The weather information system was designed to encourage multiple channels of dissemination and public sharing of weather information in order to increase the potential for active participation by farmers. Likewise, the project intended to enable as much public and free sharing and use of the weather data as possible. However, it was not clear whether and how information shared through the various distribution channels would be taken up in the villages of West Bengal. It also was not clear if, how, and why different groups of people, such as farmers, other laborers, men, and women, would learn to use the system differently. Perhaps gender or sociocultural differences have advantaged some users of the system more than others, which could enable some community members to take on roles within the system that are more important. Thus, we were interested in understanding these differences and whether the ways by which information was shared affected how and why farmers learned to participate and to what ends.

Chaudhuri, Srinivasan, and Hoysala's (chapter 4, this volume) *learning as participation* framework builds on situated learning theory, acknowledging that the learning that happens through everyday practice is a lens for understanding open practices, learning outcomes, and identity transformations. We applied parts of their framework within our investigation of the weather information system across five (out of eighteen) villages in one of the two project districts in West Bengal. We chose three villages that implemented a high number of what Chaudhuri, Srinivasan, and Hoysala (chapter 4, this volume) refer to as "manual tools" (such as an automated weather station, manual data collection units, and blackboards) along with many forms of participation (such as group meetings or discussions between project staff and villagers). The two other villages were selected for comparison, as they had either limited tools and engagement or many forms of participation but limited access to manual tools.

In this reflection, we focus on two main insights derived from our research, which highlight the advantages of Chaudhuri, Srinivasan, and Hoysala's (chapter 4, this volume) framework for connecting open practices, learning as participation, and identity shifts. The first insight concerns the benefit of differentiating between instrumental and substantive learning taking place within the system. Our attention to substantive learning helped us identify actors who were outside the scope of the intended beneficiaries of the system. For example, incorporating the concept into our research meant that we needed to change the way we were thinking about learning processes occurring within the system. We needed to consider not only how people were learning to use the system but also how participating was affecting their social position and identity (or not). The impact of the project on one group of actors within the project, the village climate kiosk volunteers (hereafter referred to as volunteers), was important and without the perspective on substantive and instrumental learning would otherwise largely have been overlooked.

However, it would not have been possible to tease out these differences had we not concentrated initially on understanding the open practices taking place within each village, including public sharing of weather information, crowdsourcing weather data, and reusing and remixing the information (as discussed in the following section). These open practices changed across villages and were also different from open processes highlighted in the literature. Likewise, the manual tools of the system were important. It was through the manual tools, such as blackboards and chalk, rain gauges, digital thermometers, and printouts, that substantive forms of learning were practiced. They served as focal points for the volunteers and allowed them to publicly demonstrate their association with the system and the forecasts and recommendations it provides. As such, these manual tools were important in establishing both the presence of volunteers and their role as *knowledgable* contributors, leading the way to the substantive learning observed. This did not happen in the village that did not have the manual tools.

The next section contextualizes the open practices taking place within the weather system, and it is followed by a story highlighting one of the volunteers, Moumita, whose name has been changed for confidentiality. This reflection ends with a summary of our position on the contributions of

Chaudhuri, Srinivasan, and Hoysala's (chapter 4, this volume) framework for open development.

Open Practices within the Weather Information System

We identified numerous ways in which *open production, open distribution,* and *open consumption* were practiced within the weather information system. Regarding *open production* practices, weather data was collected by volunteers in some of the villages. A rain gauge had been installed to measure rainfall. The volunteers had also been given a small digital device to gather daily maximum and minimum values for temperature and humidity. They collected and noted this data along with the rainfall in logbooks, which were photocopied by project field workers and brought to the project office within the district. Photocopies were then provided to the head office in Kolkata and the meteorologist to improve the forecasting models. This is an example of how crowdsourcing weather data occurred, facilitated by both the digital devices that the volunteers used and the paper-based logbooks.

Crowdsourcing weather information enabled users of the system to actively participate in improving the precision of local weather forecasts. Another benefit was that it allowed the rest of the village to understand the open production process. For example, in one of the villages, when asked to walk us through the weather collection process, a large group of men gathered and participated in describing the data collection process. This increased awareness created a sense of ownership over the data and expectations surrounding the weather information service. When observing a group meeting in another village, a female farmer accused a volunteer of keeping the manual weather data, as well as the seeds he was receiving, from the project to himself. The farmer commented, "You are growing this or that vegetable but you previously didn't, and we know that DRCSC have asked you to distribute the seed of those vegetables to our village but you didn't. Also you are not disclosing the information you are getting from those instruments.... You are not discussing it or [disseminating] it. [If it is not disseminated,] what kind of information is that?" The women's farming group was upset that the seeds were being kept from them, and the female farmer was also concerned about the lackadaisical provision of the information. They knew about the data being collected by the village volunteer and wanted him to disseminate it publicly.

When it came to *open distribution*, the system was designed to disseminate weather information via blackboards in public places and via short message service (SMS) messages. Placing blackboards with the forecasts and recommendations in multiple locations did enable completely new actors, such as kiln workers and manual day laborers, who were not part of the target population or geographic areas, to use the information. However, we also saw that only the men of the community could observe the blackboards on a daily basis (figure 7.1). Women invariably spoke of their duties that kept them busy in their homes, meaning they did not have time to go into the village square. Moreover, while men regularly spent afternoons with each other in the village squares, women tended to socialize in or around their homes. Thus, women accessed the information via their husbands or from group meetings. This limited the consumption practices of the system, as only women who participated in the village group or who had husbands amenable to providing the information could access it.

The project team also collected long lists of SMS recipients and encouraged staff to share the information with as many people as possible. Providing the information via SMS could help resolve access issues for women, presenting a reason why digital distribution channels would be beneficial in this case. However, the majority of women in this region were neither

Figure 7.1
An example of a public place mainly utilized by men.
Source: Authors.

literate nor had a personal phone. Therefore, a female villager suggested that providing messages over voice would help overcome issues of literacy, keeping in mind that mobile phones were often shared in a household. Regardless, because of technical problems, the SMS feature was not released on time, and we did not observe any use of this distribution channel.

Lastly, *open consumption* was enabled by having different measurements available for various uses and reuses of the data. For example, men mentioned using rainfall information to decide whether to irrigate their crops, whereas women used temperature information to decide whether to let their livestock graze or to let their children go to school. There were clear gendered differences in information consumption practices. These followed equally gendered divisions of labor within the villages. Men were primarily responsible for farming activities, whereas women tended to livestock, children, and kitchen gardens. While the weather information found use in all these areas, the agricultural recommendations were primarily focused on farming activities principally conducted by men.

Another important finding was that using or reusing the information was largely facilitated by human mediation. We observed how sharing information happened through discussions where volunteers and village workers interpreted needs elaborated by the men's or women's groups and then discussed what the information meant and how the villagers could use it based on their interpretations. For example, a field worker introduced the system to a new women's group by pointing out that monsoon season was coming soon and that many of them needed to work on their mud houses before it started. He highlighted how knowing whether it would rain in the next few days might help them decide when to work on their houses. By doing so, he both introduced the benefit of forecasts and reinterpreted the use of the data to suit the season and group he was addressing.

Investigating the ways in which manual and conceptual tools (Chaudhuri, Srinivasan, and Hoysala, chapter 4, this volume) were used within the villages and how open processes were both conceptualized and practiced provided valuable insights into the sociocultural settings, key actor groups, and how and why the dominant practices influenced the ways in which actor groups participated in the initiative. Interactions between the open processes and the communities of practice forming around them enabled us to dig deeper into how and why open practices and their sociocultural aspects were aiding and/or obstructing substantive learning.

Moumita's Story

In one of the villages, the blackboard is located on a main street and has been installed on the only concrete building in a hamlet that otherwise consists mainly of traditional mud houses. Moumita, an eighteen-year-old woman, comes out from her house and updates the blackboard on the street (figure 7.2). She was not originally selected to be the volunteer, as the village leader had selected her father, but her father was busy and Moumita knows Bengali. Speaking of her role, Moumita said, "I collect data twice in a day, once at 8:00 a.m. in the morning and again at 4:00 p.m. in the afternoon, and collect the data from rain gauge and temperature machine and put down the result in a copy provided to us by service center. I also write the weather data once in a week in the board. Give training to the villagers on nutrition garden with [the local village volunteer]." Beyond these tasks, Moumita's role within the village had changed because of her being involved with the project.

Moumita now had regular interactions with the village leader, or *majhi*, with whom she previously had no contact. She explained, "Since the manual collection rain gauge has been installed in the [majhi's] plot I now regularly visit his house. When it has rained I go there to collect the data and talk to [the majhi]." Moumita's father is related to the majhi, but the majhi did not previously know who Moumita was. Through her regular visits and

Figure 7.2
Moumita updating the public blackboard.
Source: Authors.

interaction with the majhi, he now not only knew who she was but had developed a positive view of her. Moumita spoke about how she was also asked by her father and other older members of the village about her new-found expertise gathering weather data and about agricultural practices in general. When asked if she has received better treatment from the villagers since becoming the volunteer, she replied: "Yes, I think my friends become jealous of me and aged people seek solution[s] regarding the weather information from me." These outcomes are especially notable considering the strongly patriarchal structure of these communities.

Moumita's story was emblematic of the changes we observed for the village workers and climate kiosk volunteers across the villages we examined. We observed volunteers becoming increasingly confident in their presentations in front of groups of men and women from their village, taking a role as a representative for the project by promoting both the weather data and other DRCSC programs. These observations indicated that the volunteers were increasingly participating in communities of practice related to interacting with DRCSC, fulfilling their project role, and networking with the village groups.

However, it is critical to note that the volunteer selection process for these roles flowed from preexisting networks of privilege. For Moumita, as the daughter of the majhi's relative, she had only been asked to take the job after the majhi had decided he saw little benefit in undertaking these activities and then asked his relative, who, in turn, handed the task to his daughter. Unlike many other girls in the village, she was literate and had completed class 12. Thus, our research confirmed how changes to the volunteers' social positions must be examined in relation to their previous positions.

Furthermore, substantive learning was only observed among volunteers and field workers. The type of learning the farmers, as the intended users of the system, engaged in was largely instrumental. Therefore, the system provided several direct benefits but did not influence the fundamental agricultural practice of the community (such as using fewer pesticides or incorporating farming practices that were more sustainable) or their social identities, positions, or relationships. Moumita's story illustrates that substantive learning requires openness that involves increased and active participation in multiple communities of practice. While instrumental learning can be beneficial, we argue that systems that only engage such forms of learning are inherently limited. Research on open practices needs to move beyond concerns of information provision, accessibility, and use toward an analysis of how they impact social positions, relationships, and identity.

Substantive Learning from Open Practices as Positive Social Transformation

The identity shifts experienced by the volunteers highlighted the relatively interconnected role of learning as participation in shaping both open practices and identity transformation. The participation of volunteers in the production and dissemination of data was a necessary element in creating the opportunity for substantive learning—an opportunity that did not exist in the village that did not offer these forms of participation. While open practices enabled both substantive and instrumental learning, the lens of Chaudhuri, Srinivasan, and Hoysala's work (chapter 4, this volume) enabled us to unpack questions surrounding *open how*, *open for whom*, and *open for what*. It allowed us to move beyond understandings that presume that free and seemingly open access either automatically or implicitly ensures accessibility or usefulness. Our focus on everyday practice as the means to identify open practices and substantive learning through ethnographic research was likewise crucial.

The concept of substantive learning as situated learning, resulting in identity shifts that could enable individuals to better negotiate their life situations, resonates well with debates surrounding social transformation through open development. Social transformation in our research context depended on the individuals, their relationships, their actions, and how and why open practices facilitated increasing connections between these elements. Desirable social transformation thus ensures substantive learning opportunities for all people, despite their gender, socioeconomic status, or social position. Practicing this type of open development implies engaging women and marginalized groups (such as the farmers) in a greater diversity of roles and providing a greater number of opportunities to participate in open production, open distribution, and open consumption practices alike. When open practices were likewise shaped and adapted to their lived experiences (through sharing information publicly on blackboards and in community farming groups instead of relying on SMS and technological solutions), there was a greater chance that people would learn to engage as full contributing members in a community of practice. We found that the emphasis on the potential for substantive learning through participating in open practices helps researchers address power relations and contextual factors that hinder desirable social transformation.

Notes

1. The Adaptation Fund was set up through the Kyoto Protocol of the United Nations Framework Convention on Climate Change (UNFCCC) as an international fund to finance climate change adaptation and mitigation.

2. India is administratively divided into states (in our case West Bengal), districts (in this case Purulia and Bankura), community development blocks or tehsils (in this case Khatna and Kashipur), and village councils (*gram panchayats*).

References

Adaptation Fund Board Secretariat. 2015. *Enhancing Adaptive Capacity and Increasing Resilience of Small and Marginal Farmers in Purulia and Bankura Districts of West Bengal.* Project IND/NIE/Agri/2014/1. Washington, DC: Adaptation Fund.

Pingali, Prabhu L. 2012. "Green Revolution: Impacts, Limits, and the Path Ahead." *Proceedings of the National Academy of Sciences of the United States of America* 109 (31): 12302–12308. https://doi.org/10.1073/pnas.0912953109.

World Bank. 2014. *Sustainable Livelihoods and Adaptation to Climate Change Project: India.* Project P132623. Washington, DC: World Bank Group.

II Coevolutionary Perspectives on Open Development

8 Understanding Divergent Outcomes in Open Development

Andy Dearden, Marion Walton, and Melissa Densmore

A Snapshot: #FeesMustFall

During October and November 2015, mass student uprisings shook South Africa as university students across the country coordinated protests to shut down all South African campuses. The #FeesMustFall movement challenged exorbitant fee increases and outsourced labor practices in an increasingly commodified higher education system. The movement's demands emphasized free, quality, and decolonized education (Naidoo 2016).

Students employed a wide range of protest tactics, such as occupying university buildings, closing access routes, shutting down classes, protest songs and marches, and extensive use of social media. Their hashtags, such as #NationalShutdown, #FeesMustFall, and #endoutsourcing (linking the students' demands to the conditions of service staff at universities), generated extensive social media activity, including South Africa's first *Tweetstorm*.

Many compared these actions to the student protests of June 1976, since the specific demand #FeesMustFall took place in the context of a broader political challenge, namely anger at the slow pace of change two decades after the advent of democracy in the country. At the same time, a resurgence of black consciousness and feminism challenged the continued dominance of white, male, middle-class norms at the university and seen in society as a whole. Black-led student movements used intersectional feminisms to challenge patriarchal and gendered practices both outside and inside the movements, which continued through 2016, culminating in another spate of protests and shutdowns in October and November of that year.

At many universities, including the University of Cape Town (UCT), face-to-face lectures were intermittently suspended in response to the protests. At UCT, for two weeks, university management and protesters were locked in negotiations. Despite reassurances from management that "the academic program is back on track and everything is back to normal," attempts to resume classes were repeatedly called off. Exam dates were set, and academic staff needed to solve the problem of how to prepare students for exams while protests continued. Many staff members proposed online education as an alternative that might avoid the risks of continuing face-to-face classes amid the protests. They suggested that recorded lectures, assessments, and tests should be made available online. It seemed an easy and obvious solution, especially at UCT, where most lectures are automatically recorded and made available to students through a web-based learning management system and high-speed campus Wi-Fi.

However, it was also apparent that this solution would exacerbate the inequities of access. Many of the poorest students did not necessarily have adequate resources and environments at home to enable them to make use of online content. Although Internet service providers quickly granted *zero-rated* access to teaching content—that is, not counting the data transfer for this material against students' data-charging plans—it failed to account for access to physical resources (Cohen 2016). Students were all too aware of these inequalities. On social media, an image of blonde-haired, white students attending a lecture, each with their own Apple laptop, went viral—just as high fees restricted access for poor students. Using blended learning was seen as advantaging the richer white students, who also appeared to be less committed to the protests. During the school year, poorer students relied heavily on the computer labs and the library on campus to access supplementary lecture materials and to do their assignments. Unfortunately, the partial shutdown of UCT also entailed a shutdown of bus transportation to campus, which made it difficult, if not impossible, for students who depended on the bus to get to campus.

After much negotiation, UCT settled on a mixed approach, with each department selecting an approach that they felt would work best for them. Departments in the science faculty offered material online but opted not to hold exams on the material, making plans to shift examinations and teaching of necessary components to the next academic year. The law faculty moved entirely online, offering laptops and mobile Internet to any student who needed them. The health sciences faculty shut down entirely, making plans to teach and hold exams on the material in a minisemester at the beginning of the following year. The outcomes of these diverse approaches varied; however, it became clear through protests against academic exclusion in March and April 2017 that the activists felt that these measures were insufficient. It should be noted that the shutdowns took place in October and November each year. These protests can be understood as a continuation of that campaign—addressing the rights of some of the student activists—but these were not at the same level as the shutdown.

In spite of an initial desire to open access to teaching materials through public sharing of educational resources online, this case demonstrates some of the ways in which this access is actually negotiated. Merely putting the material online was not sufficient. The act of *opening* teaching materials entailed addressing problems of data cost, physical resource access, and personal skills. Additionally, many students encountered emotional barriers because the stress of the situation undermined their ability to meet academic standards for inclusion or because they had strongly identified with the protest and resisted the various attempts made to smooth over the dissension.

In this chapter, we note the problematic ways in which openness is often framed, making the case that theory should address disparities in the ways these resources are accessed in practice. Increased access to digital information and the ability to forge online communities provide new ways for everyone to participate and engage. We might expect broader and more inclusive discourse demonstrating that people are debating and shaping their own futures and challenging existing power relations. However, making resources freely available does not necessarily lead directly to the hypothesized positive outcomes. We argue that research and policy should examine the processes of informal learning that occur as people engage with open services and open content to achieve their goals. We combine insights from activity theory (Engeström 1987; Kuutti 1996) and New Literacy studies (Street 1993) to develop a theoretical base for such investigations.

Introduction

Some have suggested that open development has the potential to "shift the balance of relations between haves and have-nots" (Reilly and Smith 2013, 23). Reilly and Smith argue that, unlike other flavors of ICT4D, the term *open* signifies more than access to technology. For these authors, openness involves major changes to patterns of developing and distributing information, cultural production, and knowledge in the direction of networked social morphologies. As *open* models advance, traditional hierarchically controlled models of participation make way for spaces or architectures that support networked transparency and contingency. This shift is seen as potentially enhancing human capabilities (Nussbaum 2000; Sen 1999), and that openness can help to enhance freedom; for example, by fulfilling individual rights to education or by supporting accountability (Reilly and Smith 2013).

However, there is a significant challenge for people to convert the opportunities provided by services and materials that are shared, free of charge, and openly licensed, into effective aids to achieve desired outcomes. To achieve these conversions, people must draw on not only the shared digital services and materials but also their skills, the tools that are available to them, their networks of social connections, and supporting infrastructure (in the sense of Star and Bowker 2006). Relevant infrastructure includes technical resources such as the use of Internet and web protocols, computers, mobile phones, network connections, bandwidth, and airtime. All these support (or prevent) people from accessing, interpreting, using, manipulating, adapting, and creating open digital materials. In this chapter, we use the umbrella term *resources* to refer to this diverse mix of sociotechnical infrastructure and relationships. Note that our use of the term *resources* in this chapter is broader than that used in the book's working definition of open development as "the free, networked, public sharing of digital (information and communication) *resources* toward a process of positive social transformation" (Bentley, Chib, and Smith, chapter 1, this volume). Drawing on the perspective of service-dominant logic in theories of value creation (Lusch, Vargo, and Wessels 2008), we will use the term *open digital services*, or simply *open services*, to distinguish this narrower subset of resources.[1]

People use open services in the context of practices (Bourdieu 1990) and activities (Engeström 1987; Kuutti 1996) that are meaningful in their social worlds. Here we use the term *practices* rather than, for example, skills or abilities to emphasize that people's ways of acting and behaving reflect the power relationships or structure of their societies. We also want to emphasize that they also creatively adapt such practices and exercise agency. Practices must be learned, and the adoption of open services involves the skilled integration of those new services into ongoing practices. We adopt the particular lens of activity theory (Engeström 1987; Kuutti 1996) as a way of examining these practices in context. Activity theory uses the concept of *activity* as its central unit of analysis, where an *activity* is distinguished by requiring that the unit of practice observed include a "minimal meaningful context" (Kuutti 1996, 28). This means that the unit of practice is directly meaningful for the participant (or subject) and that key elements of context, such as the tools used, the objects around which the activity is focused, and the social setting in which the activity proceeds, are all considered. We provide a more detailed introduction to activity theory in this chapter.

When open services are being incorporated into activities, most or all learning that takes place is informal learning, which happens outside formal learning environments such as schools, colleges, or training events. This learning typically is situated as people make efforts to use the services, in combination with other available resources, to address their own challenges and aspirations. For example, someone seeking to use open source software to edit media may learn by observing how someone else goes about using the software, they may make extensive use of search engines and community forums to gain understanding and overcome technical problems, and they may engage in online question and answer discussions. In these interactions, learning is motivated by the desire to edit the media, but the practices learned involve both how to edit media with this tool and how to gain support from online sources.

Such learning is unavoidably shaped by the existing resources of the learner and will be facilitated or constrained by those resources, as seen in the #FeesMustFall example. Reading technical documentation or viewing a "how to" video on a mobile phone with limited bandwidth in a foreign language is very different from the learning experience of using a desktop computer in your own language with a high-speed, uncapped connection. Similarly, the experience of asking for help, receiving responses, and interpreting those responses depends on one's background.

Fully engaging with open services involves processes of creation as well as use. To explain the imbalances between consumers and creators in open development, we draw on the field of New Literacy studies. This field suggests the notions of writing rights and writing relationships that can be used to consider the ability to design, code, and adapt open resources as well as to communicate within open services, while the notions of reading rights and reading relationships[2] describe the use of services and materials primarily as a consumer. If open development is to be truly inclusive and democratic, we must ask not only how people learn to apply open materials and services in reading relationships but also consider how services and resources can be extended to promote writing rights and writing relationships.

In this chapter, we argue that to answer the questions of whether, how, for whom, and in what circumstances the free, networked, public sharing of digital (information and communication) *resources* contributes toward (or not) a process of positive social transformation, we must attend to these informal learning processes and their realization in context. We begin by exploring situated digital interaction and situated learning in relation to open services. We consider activity theory, as framed by the work of Engeström (1987) and Kuutti (1996), as a basic theoretical framework to understand what happens when informal learning takes place. We draw attention to the role of material and social factors. An examination of how preexisting material and social inequalities are reflected when converting technical access into open services and then into desired social outcomes is also conducted. In the second half of the chapter, we extend the basic framework of activity theory by drawing on insights from New Literacy studies (Street 1993). This approach highlights differences between reading and writing relationships, asserts the importance of writing rights, and provides the concept of *literacy events*, which we suggest is a key source of data to inform the analysis of people's integration of open services within their activities.

Understanding Situated Interaction with Digital Services

Early understandings of how people use digital materials in human-computer interaction were strongly influenced by traditions of cognitive psychology. Interaction was characterized as involving perception of computer outputs, interpretation of outputs, setting goals, developing plans, and then executing those plans on the device. To make technology easier to use, designers focused on understanding the tasks that users were trying to accomplish and then mapped those to user goals, tasks, subtasks, and actions with the technology. Visual design was emphasized as a means of communicating the meaning of particular actions; for example, using action-oriented icons to create a language of interaction that would be easy for users to learn.

In the 1980s, this perspective of task-driven design was challenged by ethnographic accounts of practice as materially and socially situated. This dynamic view of practice questioned the simple division between knowledge and goals in the head of the user and information states represented by objects in the world (Hutchins 1995; Star and Bowker 2006; Suchman 1987). These more nuanced accounts of interactions between people and technology drew far more attention to the moment-by-moment arrangements of physical devices, representations, and social patterns that people used to enact their practices and achieve their objectives.

In parallel, accounts of learning saw a movement from a focus on the cognitive to paying greater attention to the material and social contexts in which learning occurs. For example, Jean Lave's ethnographic studies of apprenticeship among the Vai and Gola tailors in Liberia called into question ethnocentric views of "informal education" as something inferior to formal, schooled varieties (Lave 1997, 17). Lave's (1988) earlier account showed how the tailors' cognition was distributed across the mind, the body, and culturally organized social settings. These settings include other actors, external representations, collaborative interactions, and the materiality of infrastructure shaping access to potential collaborators.

Lave and Wenger's (1991) term *communities of practice* draws attention to these social connections as a key mechanism in many informal learning situations (see also Chaudhuri, Srinivasan, and Hoysala, chapter 4, this volume). Like other theorizations of practice, situated learning emphasizes the importance of relationships between people and their actions in the social and material worlds, and it challenges researchers to understand what motivates people to solve problems and learn as they go about their lives (Lave 1997).

Situating Informal Learning in the Context of Activities

One influential, and relatively parsimonious, framework for understanding interaction is that of activity theory (Engeström 1987; Karanasios 2014; Kuutti 1996). Activity theory approaches human practices, human-human interactions, and human-machine interactions as being always situated within an *activity*, which is the theory's central unit of analysis.

An *activity* is a behavior that is undertaken by the *subject* (person or persons) participating in the activity, that is focused by some shared *object*, which may be a tangible thing or an intangible idea, and where the subject applies *tools* to achieve a *transformation* of the object toward some outcome. For a unit of behavior to be an activity, the object and transformation must be meaningful for the subject. Activities are conducted in the context of a community that shares tools, although different members of the community may be engaging in different activities with the tools and objects. The tools and community are recognized as mediating the interactions of the subject and object within any activity. In the context of activities, an open digital service or a piece of open digital content can be considered a particular type of tool that mediates the evolution of the activity.

To understand an activity, it is important to understand its central object, which must be directly meaningful to the subject. This level of analysis focusing on what is meaningful for the subject raises the level of analysis beyond the basic task that a technical system might support. For example, a railway ticket booking system may support the task of buying a railway ticket, but the customer's activity, which is meaningful from his or her point of view, might be *arranging a weekend away*. The *object* of the activity is the weekend (or perhaps the plan for the weekend) but not the ticket. The booking system is merely one tool in this activity. This realization draws attention to other tools and actions that might influence the emerging shape of the activity, such as paper notes about the weekend plan, face-to-face discussions with the other travelers (by phone or using social media), credit cards, and hotel booking information. Activity theory uses the term *action* to refer to the more fine-grained, detailed steps in conducting the activity.

When applying activity theory to instances of informal learning, the *object* of the activities where informal learning occurs will often be neither learning itself nor the services or tools used to support that learning. If a farmer is using openly shared agricultural information, it is likely to be in the context of addressing a particular agricultural problem, and the process of using open services will be integrated with discussions with other farmers or agricultural advisers (Karanasios and Slavova 2014). Learning to use open services is a secondary action related to the activity of ensuring the healthy growth of a specific agricultural crop. Similarly, young people in South African townships who use mobile phones in creative cultural production (Noakes et al. 2014; Venter 2015) engage in extensive informal learning, but the object of their activity, as they learn, is most probably creative production, self-expression, or business development, with learning being a subsidiary outcome. In using online materials in university settings or in massive open online courseware (MOOC), the object may be to learn the subject content or to obtain a qualification, but this activity demands secondary (informal) learning processes to ascertain how to use the tools provided to enable the primary learning.

Tools in use have a shaping and constraining effect on all aspects of an activity (Hutchins 1995) and therefore on the informal learning process. This applies to physical, conceptual, or digital tools. Tools are shaped by historical and cultural factors, and they mediate the way an activity is conducted. Tools also both enable and constrain the evolution of the activity, because the transformations that are made will be those that are generally reflected in the form of the particular tool (for example, if you have a hammer, then everything looks like a nail), while other possibilities remain invisible or go unnoticed (Star and Bowker 2006).

The Social Context of Activities

In activity theory, the interplay between *community* and *subject* is recognized as being influenced by social norms and rules and by the division of labor. This structure then provides an analytic framework that can be used to explore activities in any specific context. Engeström (1987) presents the structure of an activity system using the triangular diagram shown in figure 8.1.

Division of labor is evident in the role of intermediaries (Bailur 2010; Madon 2000; Sambasivan et al. 2010) and sociotechnical infrastructure

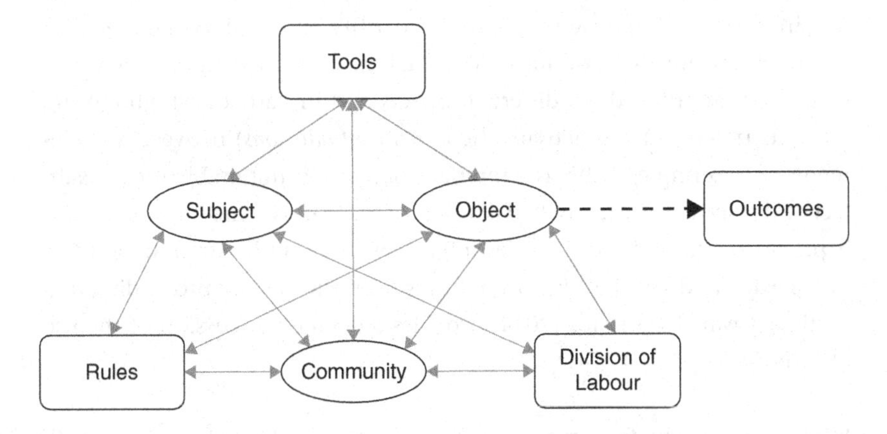

Figure 8.1
Basic structure of an *activity*.
Source: Authors, modified from Kari Kuutti, "Activity Theory and Human-Computer Interaction," in Context and Consciousness, ed. Bonnie A. Nardi (MIT Press, 1996), 28.

(Sambasivan and Smyth 2010) in enabling access and use. Intermediation can come in many forms (Sambasivan et al. 2010), including surrogate usage, where someone with technical skills accesses devices and services on behalf of a person who is less familiar with the process; proximate enabling, where a skilled user performs some part of the operation on behalf of someone who does not have the repertoire of skills; and proximate translation, where a person who is unfamiliar with the technology uses rote learning to memorize an interaction sequence they have been taught by an intermediary, for example an older person learning how to operate a digital video disc (DVD) player from a younger family member. It should be recognized that the details of how intermediations occur will be scoped by social rules influencing interactions between older and younger members of a community, between genders, or between social groups.

Intermediaries are important not only in enabling people with limited technology experience to gain initial access but also in enabling people who may have extensive technical skills in one area to make use of digital services from other domains. In the case of open government data, for example, intermediary users may be important in aggregating and packaging data in ways that make it useful to other stakeholders (Reilly and Alperin 2016). In the following section, we discuss the data journalism nonprofit OpenUp to explore the complex power relations implicit in intermediary roles.

Rules and social norms of behavior within a community are also important in the evolution of activities, so the identity of the subject and relation to the object must be examined. The analysis of individual activities proceeds by examining these different aspects and in particular highlighting tensions (referred to in activity theory as *contradictions*) between the elements. For example, if the user interface of a tool is not usable for the subject of the activity, the activity will not proceed unless some intermediation is provided. Similarly, social rules might constrain and restrict an activity, or a particular division of labor might result in the activity proceeding in a particular way. Karanasios (2014) provides a detailed discussion of the use of activity theory in ICT4D.

Interacting Activity Systems

Observing groups of people interacting with shared tools and objects may suggest that all those people are engaged in the same activity. However, this is not always the case because the meaning of an activity is bound up with the motivations, transformation processes, and the outcomes being sought by different subjects (that is, different individuals or groups). In activity theory, groups engaged in shared actions but with different motivations and desired outcomes may be understood as being involved in distinct activity systems that operate in parallel. Where this occurs, the tensions between the activities provide a useful lens for understanding the evolution of activity systems over time. This is especially true when considering that these types of digital services do not restrict, by technical design, how people may use or abuse these services, which frequently enables many types of activities to take root.

Karanasios and Allen (2013) use activity theory to examine how a diverse network of subjects, including technology suppliers, local government, international donors, local enterprises, and citizen groups, each with their own existing activity systems, came together to form a new activity system with the objective of creating a local broadband Town Information Network (TIN) in Slavutych, Ukraine. Examining the different motivations and objectives of the different stakeholders provides an account of how a new social enterprise came to be selected as the operator for the new network. This account also shows how once the network had been implemented, it was recruited as a tool, rather than as an object, in the subsequent discrete activities of different groups. For example, it was used in e-learning, e-government, and e-business. Karanasios and Slavova (2014) highlight the tensions (or contradictions) within and between the activity systems of policymakers in the Ghanaian Ministry of Food and Agriculture, district officers managing the implementation of policies, and the activity systems of field staff delivering agricultural extension services. They show how these tensions shape the emergence of the activities and actions taken at different levels.

In the #FeesMustFall example, responses to the policies of the university and the availability of online materials may be differentiated between subjects for whom the primary objective is the qualification and completion of a degree, those for whom higher education fees are the primary objective, and those for whom wider social transformations in South Africa and beyond are paramount. All these participants may be engaged in actions around #FeesMustFall, but they can be understood as participating in different parallel activities.

Divergent Outcomes Arising from Prior Inequalities

In the case of the South African university students during the #FeesMustFall protests, the intended users of the online learning materials have a shared identity, and even a shared context with shared services and materials. However, circumstances have resulted in students having limited access to the shared platforms. In this case, computing skills are roughly equal across the group, but access to economic resources required to support online education was not equal. Access to campus affected students differently: those living near campus or with access to independent transport continued to work in the labs. Fear of protesters also affected people differently. One student expressed fear of being on campus after-hours, as previous protests had featured attacks on administrative buildings and vehicles at night. In her case, as in the case of many others who often worked in the library, the situation resulted in the loss of a study environment, a problem that could not be addressed merely by increasing access to educational materials online. While many students opted to work from home, many others found their home environments were not conducive to studying. One reported that a lack of regular electricity at home meant that she could not use her computer, while another reported that her home was a small space shared by many children and other family members, which would result in other demands on her time.

Thus, the tools, which included network services and electricity, the division of labor, which included public transport drivers, and the social rules about the behavior of young women or men when at home all had an impact on the evolution of the activity. Hence, even in this relatively formal learning situation in the sense of using structured and open online learning materials, the opportunity to convert openness into benefits is revealed as being far from equal. We also know that people learn in different ways; a structured, in-person environment can be essential for some learners, while others benefit from environments that are more flexible. While the university always supported environments that were more flexible, the shift to an online approach would also negatively affect certain students. This case illustrates how open is not necessarily equal, and is actually only open for some. This result has been noted by others (Buskens 2013; Chan and Gray 2013; Graham and Hogan 2014; Gurstein 2011). It is essential that we understand the factors that mediate the usefulness of open services for all the people who might benefit from them.

Given the scale and importance of contemporary development dilemmas, a strategy focused on offering open services may well suffer from the weakness of other equal opportunity philosophies; namely, when one set of rules is applied to all, the outcomes are seldom equitable. As Lummis (2010, 44) argues, "Equality of opportunity only makes sense in a society organized as a competitive game, in which there are winners and losers." Here equal means a form of equality before the rules of the game, or equality before the law. In the game of openness, the rules involve open licenses for content or code, or content-agnostic protocols such as Transmission Control Protocol/Internet Protocol. These rules of openness are applied equally, while the people themselves are distinctly unequal. As a form of "equality of opportunity" (Benkler 2006, 19), openness is believed to provide opportunities for local innovation, spaces for newcomers to establish themselves, and chances to shift the balance of power in various ways. Yet access and use are strongly delimited by global inequities of infrastructure (Graham and Haarstad 2013).

While activity theory provides a framework for examining the situated details of interactions and informal learning, New Literacy studies draws attention to the ways in which power shapes literacy practices, including digital practices, and it examines how "activity is infused by ideology" (Hull and Schultz 2001, 588). We now consider some key concepts in New Literacy studies and how they might apply to #FeesMustFall. Following this, we turn to our second case study to understand how the ideologies of openness and the inequalities of South African society play out in the work of a nonprofit organization, OpenUp, which works in the field of data journalism.

The Practices of Literacy

New Literacy studies is an ethnographic approach to literacy that arose from an investigation of how literacy is used across different cultures (Street 1993). From this perspective, literacy involves more than just knowing how to read and write. It also involves social practices that are given meanings within specific contexts and are used to establish power relations and identities in those contexts (Barton and Hamilton 2005; Hull and Schultz 2001; Prinsloo and Breier 1996). Hence, in considering how literacy practices relate to open development and activity, our concern is with the types of reading relationships and writing relationships that emerge as people engage in activities using digital tools and platforms.

As literacy is a form of social interaction, New Literacy studies focuses on *literacy events*, or the particular social events in which writing is used. Literacy events that involve digital literacies might include, for example, *unfriending* someone on a social network site such as Facebook. Terms such as open source, comment, login, friend, and profile have all developed specific meanings. People learn these meanings not by checking a dictionary but by participating in open source communities and social networks and by engaging in the literacy events and practices of the communities to which they connect.

Specific social approaches to writing, or *literacy practices*, undergird literacy events and give them meaning in addition to ideological values in society. Thus, while middle-class parents approve of bedtime stories for young children and may encourage their children to use word processing to write their school assignments, they may be horrified by their graffiti or textisms. In the context of mobile communication, literacy practices might encompass the situationally specific meanings associated with responsiveness—what does it mean if someone is slow to respond to a flirtatious text message? Is it appropriate to ask for a job via an SMS? Thus, the definition of *literacy* is an ideological move that imposes "particular norms of social behaviour and particular relationships of power" (Buckingham 2007, 149).

The literacy practices associated with Western schooling are one form of literacy among many, but these are often given priority in open initiatives, such as academic citation conventions in Wikipedia. Anthropologists have found that these *schooled* versions of literacy may have higher social status, but they are not always as helpful to people as the literacies that are acquired and used outside school settings. For this reason, New Literacy studies focuses on people's competencies rather than their deficiencies—even people viewed as *illiterate* in school terms are often able to use literacy outside school to achieve their purposes (Prinsloo and Breier 1996). The very fact that the South African university fee protests were identified with the simple and memorable hashtag #FeesMustFall demonstrates a particular competency with communication media that is a specific literacy on the part of students organizing and promoting their case online. The acts by individuals of tweeting and retweeting around the hashtag can also be analyzed as literacy events.

Kress (1997) highlights how the right to use writing (or *writing rights*) to create highly valued *productive* writing is not universal and is far more unequally distributed than reading. Consequently, even in societies with universal access to literacy, most people's use of writing was limited to low-status reproductive texts, while productive writing was reserved for elite journalists and authors. In African contexts, Blommaert (2008) has documented the complex infrastructure needed for writing, as well as a long history of African writers struggling to establish *voice* in the face of the stigma attached to grassroots literacies.

These insights are still relevant today. Although the dominance of writing is being challenged by an explosion of multimodal digital media, new and complex infrastructure is now required to access, produce, and disseminate digital media, while the challenges of gaining voice or being heard are considerably more complex. For example, analyses of the #FeesMustFall protest data on Twitter reveal that data were dominated by the experiences of students at the historically white universities. Fewer public posts on Twitter originated from historically black universities, despite the fact that the struggles on these campuses became just as violent and frequently met more extreme brutality from police and private security (Findlay 2015). This distortion of the social media record can be explained by the same inequalities that stymied UCT's experiment with blended learning in terms of access to smartphones and data plans. Further inequalities relate to the algorithmic infrastructure of social media platforms, so although this was overwhelmingly a black-led movement, an image of white students at UCT forming a human shield to protect their fellow black students from possible police brutality became the most retweeted image of the protests during the 2015 clashes (Findlay 2015). Thus, even when asserting writing rights and enacting writing relationships in the (pseudo-) open spaces of commercial social media, preexisting material and social asymmetries result in different levels of influence in shaping the evolving conversations.

Thinking about relationships in this way includes both people and the materialities that allow them to relate to one another. The focus on writing as well as reading is a reminder of the importance of authorship as well as use in open development. Writing relationships allow participants to play a role in creating and setting goals in open development or prohibit them from doing so. Writing relationships are required if participants are to establish their position and gain command of all the *writing rights* (Kress 1997) theoretically available by engaging with open services. Our second case study, which follows, illustrates how attending to literacy events and the reading and writing relationships that surround them reveals important dimensions of the activities that are being enacted by participants in open development spaces.

OpenUp and Journalism

Data journalism finds insights from complex data sets and shares them, often in visual or story form, to make them accessible to members of the public. OpenUp[3] (formerly Code4SA) is a South African nonprofit that works to promote equal access to information and active citizenry for informed decision-making and cogovernance. It began as a team of programmers but soon expanded to include members with journalism and media backgrounds. Originally focused on technology, as reflected in the name Code4SA, the group's activities have grown to encompass a wide range of valuable online data journalism services, such as map-based interfaces to census data, visualizations of city budgets, or making the regulated prices for medicines more widely available.

Like many organizations working in open development, the ways in which OpenUp engaged with the public reflect their normative sense of openness as a social good. Thus, many of its activities seek to transform an object(s) so that it is more open. As discussed, they understood their role as intermediaries between a community of *data consumers* and *data producers* such as government and civil society. The phrase *data consumer* might suggest that OpenUp plays the role of a neutral conduit of information, but in fact the role of intermediary can involve some complex tensions between the different literacy practices, particularly in South Africa.

Making All Voices Count

OpenUp's digital literacy practices were governed by the mantra of "equal access to information,"[4] which infused web publishing with ideologies of transparency and sharing. The Black Sash, another South African nonprofit, conducts activities separate from OpenUp based around its own mission of "working towards the realisation of socio-economic rights, as outlined in the SA Constitution 1996, with emphasis on social security and social protection for the most vulnerable, particularly women and children.... We believe the implementation of socio-economic rights demands open, transparent and accountable governance (state, corporate and civil society). To this end we will promote an active civic engagement by all living in South Africa and made possible by a strong and vibrant civil society comprising community based organisations, non-governmental organisations, coalitions and movements."[5]

Thus, while the focal object of OpenUp's main activities was the information being made available online, the focus for the Black Sash was on the vibrant engagement of civil society in promoting socioeconomic rights. These parallel activity systems interacted in a coordinated community-based monitoring[6] campaign, but, in this new collaborative activity, issues of inclusive engagement gave rise to alternative interpretations of openness. While the data collected and used in the project was made available freely over the Internet, this was not truly public *in practice*, in that many members of the participating communities could not access the web; even in 2017, it was estimated that only 40 percent of South Africa's population had access to the Internet (World Wide Worx 2017). In the combined activity, OpenUp drew on the experience of the Black Sash movement and applied cross-medium storytelling practices to convey the information to the participating communities in more appropriate language. The key literacy events in the activity were creating large, colorful posters documenting how the community experienced social services, including abuses of grant recipients by the service providers, and placing them in key community venues. This approach to communication involved diverse tools and a complex division of labor between people with different skills, including data search and analysis, community activism, editorial selection of issues, and graphic design.

As the poster project suggested, achieving "equal access to information" did not always equate with conventional web publishing, and the rules of practice needed to be negotiated. Indeed, the Black Sash activists valued a different set of literacy practices, such as storytelling and interviews, which gave a more active role to the intermediary. This team had a sophisticated sense of the complexities of access in the South African context. The potential to search, analyze, and remix the information may be diminished for those who only access OpenUp's posters or mobile apps, but this is outweighed in the collaborative activity by the desire to engage older, rural, and marginalized audiences, whose information practices do not involve web use.

"Liberating" Government Data

OpenUp's Open Gazettes project highlights digital literacy practices that were potentially harmful and raises tricky questions about how the context of public data changes when it is published online. The South African government's Gazettes are official communications from the government, covering proclamations, regulations, and other official notices. Formerly, these were only available in printed format. Recently, the government made the Gazettes available via a pdf format, and a private company made a searchable digital web interface, which was available to paying subscribers. In this form, the Gazettes were used primarily by the legal profession but were relatively inaccessible to most citizens because of the inability to search through a corpus of separate pdf documents. Government departments were accused of using the obscurity of the data to conceal a controversial nuclear procurement from public debate by publishing an application for nuclear licenses in a provincial Gazette rather than in the usual national Gazette.[7] Code4SA, now called OpenUp, pointed out that the Gazettes were public documents and that the information they contained should be made accessible to the public in practice. Conceptualizing the Gazettes as a public record that documented the relationship between government and corporate interests, they set about creating Open Gazettes South Africa,[8] a free online search service, which they argued would liberate the Gazettes by supporting investigative journalistic practices and allowing searches for particular data or tracking of specific issues through an alerting service.

OpenUp used Twitter to share the new tool with followers online, and the OpenUp/Code4SA networks, their community, responded with enthusiasm, sharing a confrontational attitude toward both state and capital and excited by the prospect of future scoops. They were arguably blinkered to some extent by a shared middle-class perspective, as revealed by the tweet in figure 8.2, where the Gazette journalists showed a clear empathy with lenders rather than with borrowers. Here, the open discussion via the Twitter account, and the community that enthusiastically applauded the tool's creation, indicates some social acceptance of, or acquiescence to, this viewpoint. On the other hand, the group seems not to have considered the perspective of those thousands of South Africans whose personal details are published in the Gazette because their homes have been repossessed and auctioned through sales operated by the sheriff. Twitter posts did not

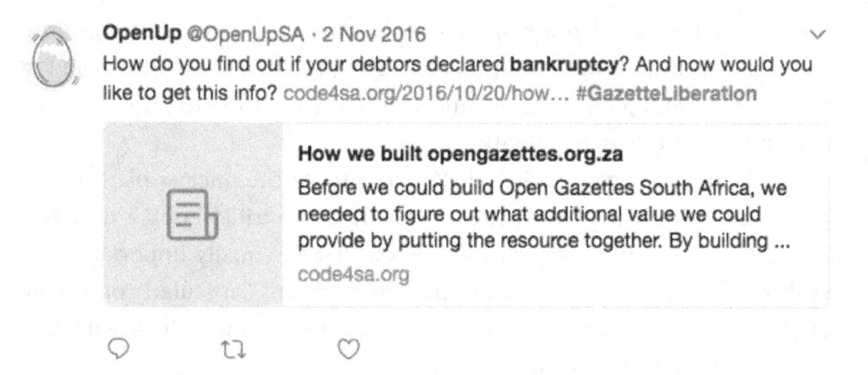

Figure 8.2
Examining the perspective of Gazette journalists.
Source: Code4SA (now OpenUp) (2016). https://twitter.com/OpenUpSA/status/793741180945268736.

succeed in connecting developers with a broader range of people, such as those with personal experience of being a distressed seller. Later, however, South Africans began phoning local radio stations, which is a literacy event where we see writing rights asserted using other tools, to complain that their identification numbers, names, and (former) addresses were being made accessible via the Google search engine.

The commercial version of the Gazette site goes as far as showing prospective real estate investors the homes via Google Street View and their locations on Google maps. This level of intrusiveness suggests the unintended potential consequences of such technologies, which draw on narrow communities and all too often inadvertently empower those with a greater share of privilege, who are more able to leverage their access to technical tools and capital to serve their own purposes.

Furthermore, individual name changes are documented in the Gazettes, including the names of women who require the spelling of their names to be corrected, as they were misspelled by the Department of Home Affairs (perhaps after marriage), and the names and identification numbers of children whose names must be changed along with those of their parents. It also contains names and addresses of spouses whose divorces are published in the Gazette. This suggests the ethical complexities and potential invasions of privacy for private individuals and, in particular, for the vulnerable individuals whose interactions with the banking and legal systems are further exposed by such a rush to transparency. This case illustrates how the social norms, or rules, of a particular group, in this case enthusiasm for using Twitter and promoting openness with less attention to privacy, influence the evolution of an activity.

While an investigative journalist's exposure of the nuclear plans could potentially allow the public better oversight over a trillion dollar deal, the distressed calls to the radio station suggest that an equally important goal of those who are claiming reading and writing rights, particularly on behalf of others, should be to adequately protect vulnerable individuals who currently do not have those powers.

OpenUp as a Community of Practice

The web developers and journalists on the OpenUp team do not learn the meanings of their literacy practices from a dictionary; they learn them in practice with collaborators and the many online communities they encounter in their activities. Where necessary, they adapt practices to support other activities that they engage in. For this reason, new members of the team are trained through apprentice-style internships, characteristic of a community of practice (Lave and Wenger 1991; Wenger 2010), and the use of their digital tools is actively promoted through journalism schools and workshops.

Gradually, in a community of practice, such new members move from peripheral forms of participation toward the center, gaining a sense of identity and belonging, as well as status and recognition, from the founders and older members. At the same time, they may undergo changes themselves as subjects within activities. Thus, new recruits are learning not only about data journalism but also about themselves in relation to other team members; for instance, the type of substantive learning that Chaudhuri, Srinivasan, and Hoysala (chapter 4, this volume) discuss. Through inclusion in the community and participation in its activities, they also acquire new literacy practices and have their writing rights affirmed.

Divergent Outcomes Arising from Prior Inequalities

This perspective of New Literacy reminds us that, just as with learning to engage with and use open services in reading relationships, learning to participate as a coproducer in open services is influenced by a person's prior history and existing repertoire of literacy practices. Two primary factors that are important in a person's decision to adopt a technology such as the service or tool and the practices associated with it are the perceived value of the technology and its perceived ease of use (Davis 1989). These perceptions are, in turn, moderated by a person's existing knowledge and experience with similar technologies. Thus, people who already have social and educational advantages may be better able to identify potential benefits of adapting their activity systems to make use of the new tools and may experience a new tool as easier to use than people who lack those advantages. Both the perceived value and the perceived ease of use of open services are further moderated by social factors. For example, if a number of peers are adopting a tool at the same time and are able to share their knowledge, this community may help individuals overcome problems they encounter, thus improving the perception of ease of use.

For middle classes around the world, the nearly ubiquitous access to ICTs and the growth of newly *conversational* forms of media have given rise to the evolution of a *participatory culture* (Jenkins 2006, 3). For example, young people from the global middle class may be exposed to fan culture that increasingly competes with and supplements formal schooling, creating educational *ecologies* distinct from classroom modes (Gee 2003; Hannaford 2016; Ito et al. 2010; Kral and Heath 2013; Sefton-Green 2006). These ecologies are strongly interconnected with child and youth culture, and media use as well as the consumption of media commodities (Buckingham 2007). A "production renaissance" driven by such media practices (Sefton-Green 2006, 296–297) involves young people participating in social media (as noted by Jenkins 2006), benefiting from nonhierarchical interest-driven opportunities for learning (Gee 2004; Gee and Morgridge 2005), and developing skills in creative production and distribution of media. These skills are related not only to media production tools but also to self-presentation in online spaces (Hannaford 2016).

Research has highlighted how the response that a person receives to her or his efforts to create a writing relationship within an open service (for example, by asking a question) is highly predictive of their future involvement in the community (Jensen, King, and Kuechler 2011; Steinmacher et al. 2015). On the other hand, the same research shows the importance of providing a narrative that legitimizes the form of the request being made, for example by indicating that the questioner is a "newbie" and perhaps using an apologetic tone. These literacy events and literacy practices of self-presentation involve skilled performances that must be learned. A person's ability to present themselves in a socially appropriate way within the community in their own language is a clear advantage (Gallagher and Savage 2015). Ludwig et al. (2014) show that command of particular symbolic and stylistic vocabularies is a predictor of engagement in online communities, and the converse of this is that those who are already marginalized by language and culture may face further marginalization within nominally open spaces. For some members of open communities, early experiences may provide a distinctive aspect of prior socialization that advantages them over others without those experiences.

Kuechler, Gilbertson, and Jensen (2012) make a comparison across a sample of open source projects showing that although 25 percent of people employed in the information technology (IT) sector are women, only 7 percent of people who post online in open source project forums at least once are women, and a mere 2.5 percent of regular posters with ten or more contributions were women. Kuechler, Gilbertson, and Jensen argue that open source projects should pay greater attention to the first social interactions that (potential) new members of the community experience. The global inequalities in representation that Graham and Haarstad (2013) observe within Wikipedia are an example of the consequences of these realities. Steinmacher et al. (2015), in turn, identify multiple barriers to participation in open source software projects, including technical hurdles, experiences of social interaction in the community, and challenges finding suitable mentors.

Building on the preceding discussion, the communicative infrastructure and *writing rights* needed to navigate, reshape, and rally nominally *open* networks are not universal and are unequally distributed. Accounts of the struggle of African writers, as observed by Blommaert (2008), to establish a *voice* in literature and the preceding discussion suggest that these struggles are also present for contemporary voices trying to access *open* services and networks from the margins. In this sense, then, full participation in open development requires multiple digital interaction resources; tools with appropriate interfaces for people to adopt within their activities; appropriate divisions of labor, including support for intermediary roles; repertoires for social behavior online; the ability to act across contexts; and the skill to transfer identities, voices, and knowledge from one context to another (Blommaert 2008; Kress and Pachler 2007). The fact that an online community provides a platform that is technically available to anyone with a suitable digital connection does not mean that the community will exhibit social characteristics that encourage diverse voices.

Conclusion

Our analysis shows how attention to the practical enactment of reading and writing relationships, through the specificity of literacy events, can be combined with other qualitative information to understand the diverse activity systems that may interact with open services. Examining how those open services are employed as tools, or objects, in a situated activity provides a rich grounding from which questions of the relationship between openness and development outcomes might be answered.

We argue that the situated informal learning of new literacies is a critical element that will be required if open services are to contribute to positive social transformation. Material inequalities in technical infrastructure and tools will constrain people's ability to convert access to open services into outcomes for themselves. Social inequalities will restrict people's ability to adjust their activities to take advantage of and contribute to new services. Social rules, social connections, and the command of literacy practices will constrain the relative ability of participants to engage with and learn to shape these spaces.

We have presented a prima facie case that "the free, networked, public sharing of digital (information and communication) *resources*" can only "contribute towards (or not) a process of positive social transformation" if attention and intervention are directed toward these material and social inequalities. We suggest that combining the theoretical frameworks of activity theory and New Literacy could enable a more detailed understanding of these factors in particular cases.

Notes

1. This perspective of services highlights the fact that the value of digital materials only arises through the use of those materials. Possessing a particular pattern of zeros and ones has no intrinsic value, but people can use those patterns to generate value.

2. We believe the idea of a reading relationship is preferable to popular notions of *skills*, *computer literacy*, or even some uses of *capability* (when viewed narrowly as an attribute of an individual person). Reading relationships inspire, nudge, shame, or force learning by potential users of open development resources.

3. See http://openup.org.za.

4. See https://openup.org.za/about.html.

5. See http://www.blacksash.org.za/index.php/about-us/about-the-black-sash.

6. Visit https://cbm.blacksash.org.za/.

7. See https://www.htxt.co.za/2016/08/24/eskom-slips-nuclear-plants-past-public/.

8. See https://opengazettes.org.za.

References

Bailur, Savita. 2010. "The Liminal Role of the Information Intermediary in Community Multimedia Centres." In *Proceedings of the 4th ACM/IEEE International Conference on Information and Communication Technologies and Development*, London, United Kingdom, December 13–16, 2010, 1–8. New York: Association for Computing Machinery.

Barton, David, and Mary Hamilton. 2005. "Literacy, Reification and the Dynamics of Social Interaction." In *Beyond Communities of Practice: Language Power and Social Context*, edited by David Barton and Karin Tusting, 14–35. Cambridge: Cambridge University Press.

Benkler, Yochai. 2006. *The Wealth of Networks: How Social Production Transforms Markets and Freedom*. New Haven, CT: Yale University Press.

Blommaert, Jan. 2008. *Grassroots Literacy: Writing, Identity and Voice in Central Africa*. London: Routledge.

Bourdieu, Pierre. 1990. *The Logic of Practice*. Translated by Richard Nice. Redwood City, CA: Stanford University Press.

Buckingham, David. 2007. *Beyond Technology: Children's Learning in the Age of Digital Culture*. Cambridge: Polity Press.

Buskens, Ineke. 2013. "Open Development Is a Freedom Song: Revealing Intent and Freeing Power." In *Open Development: Networked Innovations in International Development*, edited by Matthew L. Smith and Katherine M. A. Reilly, 327–351. Cambridge, MA: MIT Press; Ottawa: IDRC. https://idl-bnc-idrc.dspacedirect.org/bitstream/handle/10625/52348/IDL-52348.pdf?sequence=1&isAllowed=y.

Chan, Leslie, and Eve Gray. 2013. "Centering the Knowledge Peripheries through Open Access: Implications for Future Research and Discourse on Knowledge for Development." In *Open Development: Networked Innovations in International Development*, edited by Matthew L. Smith and Katherine M. A. Reilly, 197–222. Cambridge, MA: MIT Press; Ottawa: IDRC. https://idl-bnc-idrc.dspacedirect.org/bitstream/handle/10625/52348/IDL-52348.pdf?sequence=1&isAllowed=y.

Cohen, Nick. 2016. "MTN and Cell C Give Students Free Data Too." *Hypertext*, October 19. https://www.htxt.co.za/2016/10/19/mtn-and-cell-c-give-students-free-data-too/.

Davis, Fred D. 1989. "Perceived Usefulness, Perceived Ease of Use, and User Acceptance of Information Technology." *MIS Quarterly* 13 (3): 319–340.

Engeström, Yrjö. 1987. *Learning by Expanding: An Activity-Theoretical Approach to Developmental Research*. Helsinki: Orienta-Konsultit. http://lchc.ucsd.edu/mca/Paper/Engestrom/Learning-by-Expanding.pdf.

Findlay, Kyle. 2015. "The Birth of a Movement: #FeesMustFall on Twitter." *Daily Maverick*, October 30. https://www.dailymaverick.co.za/article/2015-10-30-the-birth-of-a-movement-feesmustfall-on-twitter/.

Gallagher, Silvia Elena, and Timothy Savage. 2015. "'What Is, Becomes What Is Right': A Conceptual Framework of Newcomer Legitimacy for Online Discussion Communities." *Journal of Computer-Mediated Communication* 20 (4): 400–416. https://doi.org/10.1111/jcc4.12122.

Gee, James Paul. 2003. "What Video Games Have to Teach Us about Learning and Literacy." *Computers in Entertainment—Theoretical and Practical Computer Applications in Entertainment* 1 (1).

Gee, James Paul. 2004. *Situated Language and Learning: A Critique of Traditional Schooling*. London: Routledge.

Gee, James Paul, and Tashia Morgridge. 2005. "Semiotic Social Spaces and Affinity Spaces: From the Age of Mythology to Today's Schools." In *Beyond Communities of Practice: Language Power and Social Context*, edited by David Barton and Karin Tusting, 214–232. Cambridge: Cambridge University Press.

Graham, Mark, and Håvard Haarstad. 2013. "Transparency and Development: Ethical Consumption through Web 2.0 and the Internet of Things." In *Open Development: Networked Innovations in International Development*, edited by Matthew L. Smith and Katherine M. A. Reilly, 79–112. Cambridge, MA: MIT Press; Ottawa: IDRC. https://idl-bnc-idrc.dspacedirect.org/bitstream/handle/10625/52348/IDL-52348.pdf?sequence=1&isAllowed=y.

Graham, Mark, and Bernie Hogan. 2014. *Uneven Openness: Barriers to MENA Representation on Wikipedia*. Oxford: Oxford Internet Institute.

Gurstein, Michael. 2011. "Open Data: Empowering the Empowered or Effective Data Use for Everyone?" *First Monday* 16 (2). http://firstmonday.org/article/view/3316/2764.

Hannaford, Jeanette. 2016. "Digital Worlds as Sites of Belonging for Third Culture Kids: A New Literacies Perspective." *Journal of Research in International Education* 15 (3): 253–265.

Hull, Glynda, and Katherine Schultz. 2001. "Literacy and Learning Out of School: A Review of Theory and Research." *Review of Educational Research* 71 (4): 575–611. http://www.hullresearchgroup.info/wp-content/uploads/2011/12/Literacy-and-learning-out-of-school-A-review-of-theory-and-research.pdf.

Hutchins, Edwin. 1995. *Cognition in the Wild*. Cambridge, MA: MIT Press.

Ito, Mizuko, Sonja Baumer, Matteo Bittanti, Rachel Cody, Becky Herr-Stephenson, Heather A. Horst, Patricia G. Lange. 2010. *Hanging Out, Messing Around, and Geeking Out: Kids Living and Learning with New Media*. Cambridge, MA: MIT Press.

Jenkins, Henry. 2006. *Convergence Culture: Where Old and New Media Collide*. New York: New York University Press.

Jensen, Carlos, Scott King, and Victor Kuechler. 2011. "Joining Free/Open Source Software Communities: An Analysis of Newbies' First Interactions on Project Mailing Lists." In *44th Hawaii International Conference on System Sciences (HICSS)*, Kauai, HI, January 4–7, 2011, 1–10. Danvers, MA: The Institute of Electrical and Electronics Engineers, Inc.

Karanasios, Stan. 2014. "Framing ICT4D Research Using Activity Theory: A Match between the ICT4D Field and Theory?" *Information Technologies and International Development* 10 (2): 1–17. https://itidjournal.org/index.php/itid/article/view/1213.

Karanasios, Stan, and David Allen. 2013. "ICT for Development in the Context of the Closure of Chernobyl Nuclear Power Plant: An Activity Theory Perspective." *Information Systems Journal* 23 (4): 287–306.

Karanasios, Stan, and Mira Slavova. 2014. "Legitimacy of Agriculture Extension Services: Understanding Decoupled Activities in Rural Ghana." In *European Group for Organizational Studies*, 30th EGOS Colloquium, Rotterdam, The Netherlands, July 3–5, 2014. http://eprints.whiterose.ac.uk/85166/.

Kral, Inge, and Shirley Brice Heath. 2013. "The World with Us: Sight and Sound in the 'Cultural Flows' of Informal Learning. An Indigenous Australian Case." *Learning, Culture and Social Interaction* 2 (4): 227–237.

Kress, Gunther. 1997. *Before Writing: Rethinking the Paths to Literacy*. London: Routledge.

Kress, Gunther, and Norbert Pachler. 2007. "Thinking about the 'M' in M-learning." In *Mobile Learning: Towards a Research Agenda*, edited by Norbert Pachler, 7–32. London: WLE Centre, Institute of Education.

Kuechler, Victor, Claire Gilbertson, and Carlos Jensen. 2012. "Gender Differences in Early Free and Open Source Software Joining Process." In *Open Source Systems: Long-Term Sustainability; OSS 2012*, edited by Imed Hammouda, Björn Lundell, Tommi Mikkonen, and Walt Scacchi, 78–93. IFIP Advances in Information and Communication Technology 378. Berlin: Springer. https://doi.org/10.1007/978-3-642-33442-9_6.

Kuutti, Kari. 1996. "Activity Theory as a Potential Framework for Human-Computer Interaction Research." In *Context and Consciousness: Activity Theory and Human-Computer Interaction*, edited by Bonnie A. Nardi, 17–44. Cambridge, MA: MIT Press.

Lave, Jean. 1988. *Cognition in Practice: Mind, Mathematics and Culture in Everyday Life*. Cambridge: Cambridge University Press.

Lave, Jean. 1997. "The Culture of Acquisition and the Practice of Understanding." In *Situated Cognition: Social, Semiotic and Psychological Perspectives*, edited by David Kirshner and James A. Whitson, 17–35. Mahwah, NJ: Lawrence Erlbaum Associates.

Lave, Jean, and Etienne Wenger. 1991. *Situated Learning: Legitimate Peripheral Participation*. Cambridge: Cambridge University Press.

Ludwig, Stephan, Ko De Ruyter, Dominik Mahr, Martin Wetzels, Elizbeth Brüggen, and Tom De Ruyck. 2014. "Take Their Word for It: The Symbolic Role of Linguistic Style Matches in User Communities." *MIS Quarterly* 38 (4): 1201–1217.

Lummis, C. Douglas. 2010. "Equality." In *The Development Dictionary: A Guide to Knowledge as Power*, 2nd ed., edited by Wolfgang Sachs, 38–54. London: Zed Books.

Lusch, Robert F., Stephen L. Vargo, and Gunter Wessels. 2008. "Toward a Conceptual Foundation for Service Science: Contributions from Service-Dominant Logic." *IBM Systems Journal* 47 (1): 5–14.

Madon, Shirin. 2000. "The Internet and Socio-economic Development: Exploring the Interaction." *Information Technology & People* 13 (2): 85–101.

Naidoo, Leigh-Ann. 2016. "Contemporary Student Politics in South Africa: The Rise of the Black-Led Student Movements of #RhodesMustFall and #FeesMustFall in 2015." In *Students Must Rise: Youth Struggle in South Africa before and beyond Soweto '76*, edited by Anne Heffernan and Noor Nieftagodien, 180–190. Johannesburg: Wits University Press.

Noakes, Travis, Marion Walton, Anja Venter, and Johannes Cronje. 2014. "Phone to Photoshop: Mobile Workarounds in Young People's Visual Self-Presentation Strategies." In *Proceedings of the 4th International Design, Development and Research*

Conference, Cape Town, South Africa, September 8–10, 2014, 159–180. Cape Town: Design, Development and Research Conference. http://pubs.cs.uct.ac.za/archive/00001010/.

Nussbaum, Martha. 2000. "Women's Capabilities and Social Justice." *Journal of Human Development* 1 (2): 219–247.

Prinsloo, Mastin, and Mignonne Breier, eds. 1996. *The Social Uses of Literacy: Theory and Practice in Contemporary South Africa*. Bertsham: John Benjamins.

Reilly, Katherine M. A., and Juan P. Alperin. 2016. "Intermediation in Open Development: A Knowledge Stewardship Approach." *Global Media Journal* 9 (1): 51–71.

Reilly, Katherine M. A., and Matthew L. Smith. 2013. "The Emergence of Open Development in a Network Society." In *Open Development: Networked Innovations in International Development*, edited by Matthew L. Smith and Katherine M. A. Reilly, 15–50. Cambridge, MA: MIT Press; Ottawa: IDRC. https://idl-bnc-idrc.dspacedirect.org/bitstream/handle/10625/52348/IDL-52348.pdf?sequence=1&isAllowed=y.

Sambasivan, Nithya, Ed Cutrell, Kentaro Toyama, and Bonnie Nardi. 2010. "Intermediated Technology Use in Developing Communities." In *Proceedings of the SIGCHI Conference on Human Factors in Computing Systems*, Atlanta, April 10–15, 2010, 2583–2592. New York: Association for Computing Machinery.

Sambasivan, Nithya, and Thomas Smyth. 2010. "The Human Infrastructure of ICTD." In *Proceedings of the 4th ACM/IEEE International Conference on Information and Communication Technologies and Development*, London, United Kingdom, December 13–16, 2010. New York: Association for Computing Machinery.

Sefton-Green, Julian. 2006. "Youth, Technology, and Media Cultures." *Review of Research in Education* 30 (1): 279–306.

Sen, Amartya. 1999. *Development as Freedom*. Oxford: Oxford University Press.

Star, Susan Leigh, and Geoffrey C. Bowker. 2006. "How to Infrastructure." In *The Handbook of New Media*, 2nd ed., edited by Leah A. Lievrouw and Sonia Livingstone, 230–245. London: SAGE Publications.

Steinmacher, Igor, Tayana Conte, Marco A. Gerosa, and David Redmiles. 2015. "Social Barriers Faced by Newcomers Placing Their First Contribution in Open Source Software Projects." In *Proceedings of the 18th ACM Conference on Computer Supported Cooperative Work & Social Computing*, Vancouver, Canada, March 14–18, 2015, 1379–1392. New York: Association of Computing Machinery.

Street, Brian V., ed. 1993. *Cross-cultural Approaches to Literacy*. Cambridge: Cambridge University Press.

Suchman, Lucy A. 1987. *Plans and Situated Actions: The Problem of Human-Machine Communication*. New York: Cambridge University Press.

Venter, Anja. 2015. "Smash the Black Box—Designing for Creative Mobile Machinery." In *Proceedings of the 21st International Symposium on Electronic Art 2015*, Vancouver, Canada, August 14–19, 2015, 1–8. Vancouver: New Forms Art Press, artists and writers. https://isea2015.org/proceeding/submissions/ISEA2015_submission_203.pdf.

Wenger, Etienne. 2010. "Communities of Practice and Social Learning Systems: The Career of a Concept." In *Social Learning Systems and Communities of Practice*, edited by Chris Blackmore, 179–198. London: Springer.

World Wide Worx. 2017. *Internet Access in South Africa 2017: Executive Summary*. June 19, 2017. http://www.worldwideworx.com/wp-content/uploads/2017/07/Exec-Summary-Internet-Access-in-SA-2017.pdf.

9 A Critical Capability Approach to Evaluate Open Development

Yingqin Zheng and Bernd Carsten Stahl, with contributions from Becky Faith

Introduction

The concept of *open development* has evolved over time. Smith et al. (2008) presented Open ICT4D as a working hypothesis that refers to the use of new ICTs to engage in *open* processes to achieve development gains. It is underlined by a set of principles that privilege universal access, participation, and collaboration over restricted access and centralized production (Smith and Elder 2010). Stemming from ICT4D roots, the openness of concern in this context is considered to be digitally or information-network enabled. Open development refers not only to the technological infrastructure that supports openness but also more importantly to the processes that enhance sharing, collaboration, social inclusion, accountability, and transparency.

In recent years, open movements have spread across many sectors, including open data, open government, open educational resources, open science, and open development, and standards for openness have been adopted in many of these sectors. In June 2013, the European Union adopted an updated directive aimed at facilitating the reuse of public sector information by businesses, creative citizens, developers, and others (European Commission 2013). At the same time, the G8 countries signed the Open Data Charter, built around five principles: open data by default, high quality and quantity, usable by all, release data for improved governance, and release data for innovation (Susha et al. 2014). This in itself could be seen as demonstrating the importance of digitally enabled openness to a range of international and national political entities both in middle-income countries and in the Global North. In many cases, openness is being promoted by institutional players such as international organizations, national and local governments, and nongovernmental organizations.

Given the spread of these initiatives, it is important to critically examine the extent to which openness serves the interest of people and organizations in less advantaged positions vis-à-vis those with access to greater resources and network power. One thing we have discovered is that these initiatives are often backed by the normative assumption in the literature that openness is good in and of itself. For example, open government is considered able to improve governance by increasing transparency, participatory decision-making, and project development (Cyranek 2014). Claims are made regarding the potential development benefits of open science despite uncertainty about the outcomes (Chan, Okune, and Sambuli 2015).

The literature on open development, however, shows a distinct absence of evidence of impact (Bentley and Chib 2016). This is most likely caused by multiple factors. One factor is that many of these open development initiatives are relatively new, and thus their impacts might not yet be felt. Another is that there are many initiatives that have not succeeded. A final reason is that there is a lack of evaluative frameworks on the impact of openness in development, which could be good and bad, inclusive and exclusive, and so on.

As with other information and communications technologies for development (ICT4D) initiatives, open development faces challenges such as entrenched power relations, institutional resistance, and a lack of skills. The societal impact of open development is also challenged by asymmetries in digital infrastructure and access to supportive resources on the global and local levels. It would be simplistic to assume that openness automatically brings social benefits (Gurstein 2011; Smith and Reilly 2013). As Buskens (2011, 71) points out, "Open ICT ecosystems do not exist in a power vacuum; neither does our (nor anybody else's) thinking about open development." Indeed, diffusing technologies in societies without addressing underlying inequalities resulting from gender, class, and various forms of social barriers may serve to reinforce inequalities and consolidate segregation. For example, without sufficient technical and institutional protections to safeguard individual privacy and interests, participating and sharing on open platforms may render users inadvertent victims of commercial interests seeking profits from big data (Boyd and Crawford 2012). If we wish to understand the connections between openness and impacts, we need effective evaluative tools so we know what these impacts are. In the next section, we lay out one such approach.

A Critical Capability Approach to Open Development

How do we evaluate the impact and implications of open development in societies while accounting for structural, ideological, and political influences? If open development researchers are serious about making a contribution to human development, it is important to take the *development* part seriously rather than taking it for granted, to acknowledge the normative nature of their work, and to explicitly reflect on the values behind the prescriptive and evaluative research. Open development research can benefit from engaging with development discourses, not only mainstream economic theories but also difficult questions related to justice, equity, culture, and universal values.

In some instances, for example, open development contests existing value systems and power relations with uncertain consequences. An example is the global intellectual property regime established by developed economies versus alternative forms of property rights regimes promoted by the open source movement, which raises questions about distributive justice (Rawls 1971) in development ethics (Reilly 2014). Open educational resources (OER) are challenging publishers' dominance in the content business with much pushback. Some are also slowly morphing into educational platform providers, bringing a very different level of control over educational content, data, and other information. Meanwhile, the emerging global technological standards, such as open data standards, are still often set by the Global North, while poorer nations are invariably followers. To what extent does openness challenge or build technological hegemony or path dependence? How can the voices and interests of developing societies that lack the technological know-how be included?

There are rich intellectual traditions in areas such as philosophy of technology, critical theory, science and technology studies, and critical perspectives of information systems (Stahl 2011) that provide insights into the role of technology in society. We draw on the critical theory tradition and development ethics to develop the critical capability approach (CCA) (Zheng and Stahl 2012) as the basis of a design and evaluative framework. The evaluative framework consists of sets of research questions for the consideration of researchers and practitioners with regard to the design, implementation, and evaluation of open development initiatives. In the rest of this section, we will briefly explain the conceptual basis and the four principles of the CCA.

There are two main sources for the intellectual tradition that underlies the CCA: the critical theory of technology and information systems and Sen's capability approach (CA). The critical theory of technology (CTT) argues that technology is socially constructed or shaped and that its development strongly influenced the beliefs and interests of various social groups (Feenberg 1999). Rather than a mere tool, technology is considered to embody values and preferences, which could exert significant structural power over society when rules and patterns of behavior are inscribed in technical design and become blackboxed, institutionalized in rules, regulations, and practices, and no longer questioned or scrutinized. The affordances of technology depend greatly on the socioeconomic environment, institutional settings, and social practices. Moreover, openness has the potential to better social conditions and even promote emancipation but often has intended or unintended consequences resulting in the opposite effect.

Sen's CA is an important framework of thought in the discourse of human development and often serves as a critical alternative to mainstream ideologies such as economic growth and modernization (Sen 1984; Sen 1999; Alkire 2005). The CCA is an operationalization of Sen's CA that integrates the core concepts from Sen with those from critical theory. There are many ways to operationalize Sen's CA (Robeyns 2005; Robeyns 2006). Examples of applying the CA to technological development include technology as empowerment (Johnstone 2007), the conceptualization of inequality in digital exclusion as capability deprivation (Zheng and Walsham 2008), ICT4D in general (Zheng 2009), human-centered design (Oosterlaken 2011), and the choice framework (Kleine 2010). The latter is based on the conceptualization of *choice*, external to the original vocabulary of the CA but derived and extended from the concept of *capabilities*. In comparison, the CCA (Zheng and Stahl 2012) draws on the original core concepts from Sen, focuses on its basic principles, and integrates them with a critical conception of technology.

From a CA perspective, ICT is typically seen as being embedded in the process of human development, and its purpose is to enable individuals to lead a life they have reason to value (Zheng 2009; Oosterlaken 2011). The evaluation of technology and technological processes is thus based not on utilities but on the expansion of an individual's level of well-being, available opportunities, and the range of choices under their specific social circumstances. With an emphasis on *agency freedom* (Crocker and Robeyns

2009), the CA urges researchers to consider the agential capacity of ICT users beyond that of being passive receivers of innovation. This resonates with the notion of open development, where participants are considered both users and producers of information and knowledge. Incorporating individual agency in studying open development sensitizes us to the need for public discussions, participation, and social inclusion in the process of opening up, sharing, and utilizing information resources.

The CCA can be used to assess the motivations driving open development initiatives and their social consequences. The CCA consists of the following four principles (each of which will be discussed briefly):

1. The principle of human-centered development
2. The principle of human diversity
3. The principle of protecting human agency
4. The principle of democratic discourses

The Principle of Human-Centered Development

The CCA suggests that digital development should be based on the imperative to enhance substantive freedom that will enable individuals to lead a life they have reason to value and to remove injustices and barriers to freedom. Neither the CA nor CTT specifies, or believes that it is possible or desirable to specify, an ideal society. Both therefore focus on generating opportunities and removing barriers to freedom and justice in specific contexts. The principle of human-centered technological development stands for the need to reflect on whether open development actually enhances individual capabilities. For example, to what extent does open government really improve public service and transparency, and for whom? Also, under what circumstances does open education improve the experience and result of education?

By insisting that any kind of open development should be human centered, the CCA raises the important question of whether open initiatives have or could become hegemony in themselves, whether the potential consequences have been explicated and deliberated, and whether alternatives have been or could be explored. For example, standards in open data and open government benchmarking are value-laden without sufficient considerations of the challenges of implementation and organizational transformation (Susha et al. 2014). Some scholars are wary of the dominance of

Western culture's normative values, which may lead to the suppression of local culture and local value systems (Flor 2015; Gregson et al. 2015).

The Principle of Human Diversity

The second principle of the CCA is that of human diversity, ranging from individual characteristics, to environmental conditions, to social arrangements (Robeyns 2005). Open development should therefore be evaluated not on a singular dimension but on the plurality of functionings and capabilities (Sen 1993). The concepts of functionings and capabilities are explained as follows: "A functioning is an achievement, whereas a capability is the ability to achieve. Functionings are, in a sense, more directly related to living conditions, since they are different aspects of living conditions. Capabilities, in contrast, are notions of freedom, in the positive sense: what real opportunities you have regarding the life you may lead" (Sen 1987, 36).

As Robeyns (2005) shows, the realization of capabilities is inevitably subject to personal, social, and environmental *conversion factors*. Personal characteristics include mental and physical conditions, literacy, and gender; social factors could be social norms (for example, the role of women, rules of behavior, materialism, and religion), institutions (for example, the rule of law, political rights, and public policies), and power structures (for example, hierarchy and politics). Environmental characteristics, such as climate, infrastructure, and public goods, also play a role in the conversion from goods and resources to capabilities.

As people are endowed with various physical and mental characteristics and live in diverse environments under different sociocultural conditions, these factors will affect what real opportunities a person can realistically enjoy by adopting a particular technological arrangement. For example, a low level of information literacy and digital skills and the absence of legal and policy frameworks are major barriers for open development in resource-poor environments. Consequently, infomediaries (Linders 2013) and the inclusion of multiple stakeholders in a dynamic *open data ecosystem* are important (Smith and Elder 2010).

In short, the principle of human diversity invalidates any assumption about the universal benefit of openness. Rather, when evaluating openness, it is necessary to ask what capabilities different types of open initiatives expand, for whom, under what circumstances, and what the enabling factors and the barriers are.

The Principle of Protecting Human Agency

The concept of enhancing human agency is closely linked to *emancipation*, which is at the core of critical theory (Stahl 2008). An important implication is the resistance to reification, the process of rendering a socially malleable phenomenon as an apparently *objective* thing. For example, openness has often been reified to represent efficiency and progress, while, in practice, technological adoption and diffusion have been inevitably embedded in dynamic social processes, giving rise to emergent and often unintended consequences (Avgerou 2010). For example, the proliferation of social media, crowdsourcing, and big data offers significant transformative opportunities yet incurs serious concerns, such as the erosion of individual privacy and autonomy, large-scale surveillance, and potential discrimination. These concerns are often unquestioned or even accepted as normal (Cilliers and Flowerday 2014; Sen et al. 2013; Zuboff 2015). The implications for invasion of privacy or knowledge integrity, or the boost to the institutionalized power of surveillance, may not be fully appreciated before individuals and organizations are persuaded to be open and sharing.

The reification of a particular technological or institutional arrangement can obfuscate choices that users could have, thus potentially limiting agency and freedom. The CCA takes the position that people should always retain the power as an autonomous being to make decisions on their choices in life, whether they relate to the adoption and adaptation of information resources or a particular model of development (Crocker 2008).

Some argue that openness may enhance users' choices by making available the *freedoms* for users to produce, use, reuse, and remix data, thereby raising the agency level. For example, groups who have been historically excluded from scientific inquiry might be able to participate in open and collaborative science, thus arguably increasing their autonomy, agency, and chances in life (Chan, Okune, and Sambuli 2015). On the other hand, individuals may have more or less agency than others in the context of open development. Given the existing asymmetry of power and access to resources within and across societies, it could be the case that knowledge sharing is used to serve the interests of private business or those in advantageous positions.

An important part of the open movement is empowering individuals not just through open dissemination of data—that is, as the recipient of information—but also through opportunities for the public to exercise their agency to serve their own objectives, for example in monitoring and providing feedback to the government, innovating and providing public services, or in knowledge creation and scientific collaboration (Corrêa et al. 2014; Linders 2012; Phang, Kankanhalli, and Huang 2014; Sandoval-Almazán 2015). The literature shows that most open development initiatives are still at the stage of releasing and sharing data and encouraging usage of data, and are rather limited in achieving transformative effects in public participation, citizen empowerment, and collaborative innovation, especially in the case of open government data (Davies and Perini 2016; Janssen, Charalabidis, and Zuiderwijk 2012). Peixoto and Fox (2016) suggest that while ICT platforms may improve the capacity of policymakers and senior managers to respond, there is not necessarily a willingness to do so. Nevertheless, there are also successful cases of user participation in crowdmapping or participatory mapping (Panek 2015), such as Ushahidi and Open Street Map.

The Principle of Democratic Discourses

Emancipation proceeds by revealing the sources and causes of the distorting influences that hide alternative ways of life from us (Hirchheim, Klein, and Lyytinen 1995, 83). For developing societies, being open to possible alternative development paths is of particular significance, especially when the global order and development models have been largely defined by the Global North and often imposed on poor nations.

Both critical theory and the CA consider democratic discourses as being pivotal to any social changes. ICTs, and increasingly the functions of the production, storage, and access of data, have become a core infrastructure of modern societies. It is therefore worrisome that, unlike most other aspects of modern societies, these developments are too often driven by a relatively small number of sectors, such as technological developers, international organizations, and government agencies, and largely removed from democratic control. Sen follows John Rawls in perceiving democracy as "the exercise of public reason" (Crocker 2008, 307) rather than restricting it to the procedure of majority rule, but more importantly democracy "requires the protection of liberties and freedoms, respect for legal entitlements, and the guaranteeing of free discussion and uncensored distribution of news and fair comment" (Sen 1999, 9–10).

The importance of democratic discourses for the future of digital development has long been recognized by philosophers of technology (for example, Brey 2008). However, implementing democratic control of technology raises numerous problems, such as issues of property and distribution. For example, which democratic structures would be required for shaping emerging technologies or social arrangements based on open development initiatives? A fundamental question is how to evaluate open development and how to structure democratic discussions about it. At the minimum, a participatory approach to openness will require a continuous process of monitoring and debating. It will also require different ownership and control models of open and shared data, which need to be opened up to deliberation by all stakeholders (Crawford and Schultz 2014).

Evaluative Framework

The CCA principles serve as a conceptual basis for reflecting on policies for open development. By explicating the ideological assumptions and political processes behind open development policies, they create space for broader discourses on development processes—for example, by allowing alternative or complementary creative solutions tailored to local contexts to be explored—and demarginalize other important aspects of social development, such as institutional and cultural processes.

In this section, we suggest an evaluative framework that operationalizes the CCA principles to form the basis for empirical research, consisting of a set of evaluation questions. These evaluation questions are structured in terms of *design, implementation,* and *evaluation* of open development projects to ensure that the entire project life cycle is subject to critical reflection. In regard to each of the three phases, we start with a general discussion about key issues needing investigation. We then provide a table of questions that can orient the research and give input to the researchers when designing a specific study.

Design: Structural Context of Openness

The first step in operationalizing the CCA as a means of investigating open development initiatives is to examine their starting point. The aim is to reveal the political interests that are promoting openness in a particular setting, and the standards and structures they are putting in place. This could be done by looking at who the drivers of the openness project are, the definition of openness used in the project, and whether the project attempts to meet a preagreed set of openness standards. There are various international standards for openness, created by donors, international organizations, and governments. It is important to note which, if any, are being adopted or adapted to local contexts. It is also useful to understand whether projects are promoting openness for the sake of being open or for other reasons.

The concept of openness could be focused on Western culture and clash with other cultures and belief systems. Flor (2015) discusses a case in the Philippines where an open knowledge system to capture best practices in climate change adaptation among Indigenous tribes was met with resistance because it clashed with the Indigenous belief system. The Indigenous communities, as a rule, have tribal elders, chieftains, and healers, who regard themselves as custodians of knowledge, which may only be shared with prudence, responsibility, and, on occasion, sanctity. Wen and Liu (2016) also found that the principles of open knowledge that embrace open expression and cooperative learning clashed with the educational culture in Taiwan, which emphasizes obedience to authority. Rao et al. (chapter 4, this volume) discuss how OER may be a source of intellectual hegemony, which calls into question whether OER content can be trusted in developing contexts.

If the project is using a preagreed definition of openness, then it is important to understand the rationale behind this definition: whether claims are being made regarding the potential of openness to save money, increase transparency, or boost participation. Gagliardone (2014, 3), for example, suggests that in Ethiopia the government made claims about the openness of an ICT for development program while appropriating donors' demands for openness and democratization "to support its ambitious state- and nation-building process."

Ownership of the definition of openness is also important: whether the definition is being imposed on beneficiaries or whether they are able to modify it. The issue of power structure is thus fundamental. Johnson (2014, 270) points out that the opening of data "can function as a tool of disciplinary power ... [that] enhances the capacity of disciplinary systems to observe and evaluate institutions and individuals' conformity to norms that become the core values and assumptions of the institutional system whether or not they reflect the circumstances of those institutions and individuals."

For the participants of the project, it is important to understand the incentives for participation (Miller 2009) and whether the sharing of data and information gives rise to inequality and exclusion. These issues relate to the broader power dynamics of the project and the interests it is aiming to promote, who has power to make decisions about the project, and deciding how it is implemented. Gregson et al. (2015, 42) also warn that local and Indigenous knowledge and voices that are already marginalized by dominant Western-centric discourses and epistemes might potentially be further suppressed by open knowledge platforms. Similarly, Graham et al. (2014, 749) studied the uneven geographies demonstrated on Wikipedia and suggest that "despite Wikipedia's structural openness, there are fears that some parts of the world will be heavily represented on the platform and others will be largely left out, a situation that could simply reproduce worldviews and knowledge created in the Global North at the expense of southern viewpoints." Furthermore, gender issues are largely ignored in the literature (Bentley and Chib 2016), although an interesting study by Terrell et al. (2017) points out that bias against women persists in open source communities. Table 9.1 provides a suggested list of questions for consideration during the design stage of open development projects.

Implementation: Enabling Factors and Barriers

The next set of issues concerns the outcomes of openness: how they are benefiting beneficiaries and whether the outcomes are expanding people's choices and participation in political processes. These relate to the

Table 9.1
Questions concerning the design stage of open development projects

1. What is the definition of openness used in the project? What is the rationale behind the definition of openness used? Who has participated in the creation of this definition? Who has been able to modify it at all?
2. Is the project attempting to meet a preagreed set of openness standards? If so, what are those standards?
3. What are the intended consequences of the project? Also, what is your (the researcher's) understanding of whose interests this project intends to promote?
4. Who has power in this project? For example, who is making decisions about this project and deciding how it is implemented?
5. What are the incentives for participation, and do users have reasonable alternative options?
6. What are the foreseeable but unintended consequences, and how are they being addressed?
7. Who are the stakeholders involved, and what are their roles?

principles of human diversity and protecting human agency. The objective is to identify the social and institutional arrangements and circumstances necessary for openness to truly bring benefits (table 9.2). Barriers to openness might be political or take the form of corporate interests or the risks of promoting openness in a politically closed environment (Njihia and Merali 2013). There might also be ethical considerations associated with promoting openness or transparency in ways that might undermine the interests of disadvantaged or marginalized communities. There are also risks that resources made open could fall prey to commercial forces.

We need to examine what conversion factors are in place to support openness that is truly beneficial. Examples of conversion factors include (1) a legal framework; (2) a supportive political culture; (3) people with the right skills and commitment to the project; and (4) physical infrastructure such as sufficient bandwidth and hardware.

Information infrastructure has always been identified as one of the basic building blocks of digitally enabled social change, and the same applies to open development. In Iraq, for example, Al-Taie and Kadry (2013) identify the absence of effective ICT infrastructure, the lack of ICT literacy among teachers, slow and expensive Internet access, and an inadequate power

Table 9.2
Questions regarding the implementation stage of open development projects

1. Who are the beneficiaries? How is openness benefiting beneficiaries? Has the implementation been approached from the perspective of the beneficiaries? Does the beneficiary group have an understanding of openness? Have the project implementers done user-focused work looking at this issue?

2. What is the impact the project aims to achieve by being open? How is openness supposed to support organizational goals and the broader development processes?

3. What are the preconditions, and what has to exist in order for openness to be truly beneficial? What social and institutional arrangements and circumstances are necessary for openness to truly bring benefits?

4. What are the barriers? For example, how open is the government already, and are you going to be putting yourself in danger by promoting openness in a closed regime? Or are there threats from corporate interests?

5. What are the ethical considerations? For example, how might openness undermine people's safety or interests?

6. Who are included in the open initiatives and who are excluded? What are their roles? What data is included and excluded? Should there be boundaries for openness and, if so, where and how should they be drawn? What kinds of regulations and structures are in place to protect user data and privacy?

supply as just some of the barriers in open data processes. Gregson et al. (2015) point out that in many regions in Africa and Central Asia, essential technological structures to facilitate open access and knowledge sharing are lacking, thereby limiting visibility and the international impact of scientific publications from these countries. In areas with underdeveloped infrastructure, it might therefore be necessary to combine digital technology with other appropriate media that suit the context and user capacity to broaden user access to shared resources or to participate in open government processes. Other relevant factors may include the skills of citizens to access and use data, such as language, information literacy, and proficiency in domain knowledge (Graham et al. 2014; Martin et al. 2014). Graham and Hogan (2014) found that editing articles on Wikipedia is an intensive task "replete with a great deal of barriers to entry," hence the dominance of representation from the Global North.

Legal frameworks, such as an open data commons, are important in stipulating ownership and rights in relation to public data, as well as to ensure trustworthiness of data (Peled 2013). For example, Norway has been known for providing appropriate legislation, technical standards, and architecture to ensure the completeness, accuracy, timeliness, and reliability of public data (Thurston 2012). For open initiatives focused on transparency and accountability to have any effect, the legislation of rights to information needs to be in place (Trapnell 2014) and supported by other legislation that protects citizens' freedom of expression and privacy. Such legislation also needs to be embedded in a receptive political culture. In Uganda, despite the enactment of the Right to Access to Information Act in 2005, a hostile legislative framework, including a Regulation of Interception of Communications Act (2010) and a Public Order Management Act (2013), has challenged advocacy efforts toward public accountability, transparency, and citizen participation in governance issues (Baguma 2014).

It is also illuminating to look at who and what is included or excluded in open initiatives. For example, certain data might be withheld and certain groups excluded. This exclusion can be intentional or unintentional. If it is intentional, we should explore the reasons why the project is designed or structured in such a way. Alternatively, unintentional exclusion might be caused by existing hierarchical power structures. McLennan (2016), in a study about a peer-to-peer platform in development, observed that while the network bypassed traditional intermediaries and promoted diversity, it became dominated by American participants, and the network founder and key participants created new intermediaries. The transfer of the network to Facebook also led to further exclusion of participants in Honduras.

Furthermore, researchers should also examine whether boundaries for openness should be designed, such as inclusion or exclusion of content, participants, and access control. Kleine, Hollow, and Poveda (2014) point out that opening health data can pose significant privacy issues related to collecting, storing, and analyzing medical data. Thus, notwithstanding the potential public benefit, one must consider where and how boundaries should be drawn and what kinds of regulations and structures are in place to protect user data and privacy.

Evaluation: Who Has a Say and Who Benefits?

The final set of issues concerns the evaluation of open development: exploring the kinds of democratic structures required to make decisions about open development related to the principle of democratic discourse. First, these concern the questions of how the project is evaluated and the indicators for success, such as who is setting the indicators and what the sanction is for not meeting them. We then need to identify the (claimed and actual) beneficiaries of the project and how openness is benefiting them. It is possible that intended beneficiaries may not have a clear understanding of openness, instead feeling that it is being imposed on them or provided for them. From the perspective of Sen's CA, researchers should examine the *actual* project outcomes in terms of beneficiaries' capabilities (Zheng and Walsham 2008; Zheng 2009). These capabilities might include development impacts such as improvements in health, financial situation, or sanitation, or enhancing individual autonomy, voice, creativity, and participation. This raises additional questions: What can they do as a result of the project that they could not do without it, and what value does it provide them? What is the impact the project aims to achieve by being open? Are these objectives met? How does being open support the broader development program and organizational goals? These questions are elaborated further in table 9.3.

Many open development projects work with communities that can be very large, national, or even global, with uncertain boundaries. This inspires new questions such as: Who draws the boundaries, if any? If information made open for one community is too technical or complicated for ordinary users but easily captured by individuals and organizations with resources and capacity,

Table 9.3
Questions concerning the evaluation stage of open development projects

1. How is the project evaluated? What are designated indicators for success, who is setting the indicators, and what is the sanction for not meeting them?
2. What are the outcomes of this project in terms of peoples' well-being—what capabilities does it contribute? What can they do as a result of the project that they couldn't do earlier, and what value does the project provide for them?
3. What are the outcomes of the project in terms of increasing people's choices and participation, for example, in political processes?
4. What are the mechanisms for incorporating feedback and improving services? Who are the stakeholders and how are they involved in discussions?
5. If the organization is committed to having openness persist, are there related long-term indicators and a strategy for accomplishing this?

does it enhance inequality as a result? For example, Peled (2013) points out that the open data initiative based in the United States benefited a limited set of stakeholders and empowered those who already possessed the funds and expertise to use data (such as corporations and software developers).

Mechanisms for incorporating feedback and improving services should be examined to see how stakeholders are involved in discussions. Research should also seek to illuminate who has the power in these discussions: which stakeholders have the power to promote and engage with openness, and who is excluded from these discussions. If organizations are claiming to be committed to openness for the long term, it would be interesting to explore whether these claims are backed up by long-term indicators and strategies.

Evaluative Questions for Different Types of Open Development Projects

So far, the evaluative framework does not take into account different types of open development projects. Smith and Seward (2017) suggest that openness involves three distinct processes: open production, open distribution, and open consumption. Open projects may involve only one or two processes or all three of them. Smith and Seward's typology provides guidance to broadly identify different types of open development projects and associated relevant processes and practices. It follows that, to evaluate these different projects, emphasis may be placed on different questions. Table 9.4 includes some questions based on the categorization proposed by Smith and Seward (2017). The questions are again suggestive rather than exhaustive, and they are designed

Table 9.4
Three main types of open processes

Open process	Open practice	Suggested relevant questions
Open production	Peer production E.g., open source software production, Wikipedia, open legislation	Who produces open resources, and who are the intended users? Who decides what to produce? Who can participate, how, and to what extent? Who is included or excluded from open production?
	Crowdsourcing E.g., open innovation, citizen science, Ushahidi, ICT-enabled citizen voice	How do the rules, norms, and values of the process affect participation? How does the software or technology affect participation? What standards or formats are adopted? Who defines them? How is the accuracy and quality of shared information resources ensured? Could shared information resources be used in ways that disadvantage vulnerable groups?
Open distribution	Sharing, republishing E.g., open government data portal, OER portal (e.g., Khan Academy), open access journals	To what extent does the format of the shared resource render it easily readable and transferable? In what ways, and for whom, does the licensing regime for shared information resources affect the uptake of open resources? Are there mechanisms to enforce licensing regulations across different countries or communities? How are privacy and security concerns, if any, addressed when using and sharing certain data? Could shared information resources be used in ways that disadvantage vulnerable groups?

Table 9.4 (continued)

Open process	Open practice	Suggested relevant questions
Open consumption	Retaining, reusing, revising, remixing E.g., translating educational materials, taking a massive open online course (MOOC), intermediary visualizing open government data	What skills and capabilities are required to take advantage of open information resources? What technological capacities are required? What other conditions are needed to facilitate open consumption (e.g., technological infrastructure, intermediaries, and training)? Who is engaging in open consumption, how, and to what end? How does consumption of open information resources contribute to people's well-being and opportunities? To what extent does open consumption improve an individual's agency, their ability to change their lives or participate in public life? Who benefits from the output of open production and who does not? Does it unfairly advance the interests of certain groups at the expense of others? Could shared information resources be used in ways that disadvantage vulnerable groups?

Source: Smith and Seward (2017).

to generate reflection and further discussion by researchers and practitioners when they interact with different open processes and practices. Some questions may overlap, as they are relevant to multiple types of projects.

Research Methodology

Critical research in general is interventionist. The purpose of the research is not just to describe a phenomenon but also to improve the state of the world, to make a positive difference, which is often described in terms of promoting emancipation. This is a difficult goal to achieve, and it is also often difficult to measure the degree to which it has been successful.

In order to establish a critical consciousness in operationalizing the CA, Klein (2009) suggests the following six criteria be applied in empirical research that seeks to provide a critical evaluation and discussion on open development:

1. Being concerned with the conditions of human existence that facilitate the realization of human needs and potentials;
2. Supporting a process of critical self-reflection and associated self-transformation;
3. Being sensitive to a broader set of institutional issues relating particularly to social justice, due process, and human freedom;
4. Incorporating explicit principles of evidence given (or an explicit truth theory) for the evaluation of claims made throughout the research process;
5. Incorporating principles of fallibility and self-correction (growth of knowledge through criticism; for instance, the principle of fallibilism);
6. Suggesting how the critique of social conditions or practices could be met (as a safeguard against unrealistic and destructive negativism).

There is no single methodology to be employed by critical researchers. This is true for the use of the CCA to investigate open development projects. The outline and research questions that were presented can be implemented using various methodologies. When choosing the details of the methodology and developing a research plan, researchers should ask themselves whether they want to contribute to the critical goal of making a difference and which methodology is likely to contribute to that goal.

There are some research approaches and methodologies that explicitly embrace an interventionist approach and that are therefore suitable for critical research. The key methodology used for this purpose within the field of information systems is action research (Avison, Baskerville, and Myers 2001; Baskerville and Wood-Harper 1998; Mumford 2001). The form of action research that most explicitly aims to include affected and often underrepresented communities is participatory action research (Ahari et al. 2012; Argyris and Schön 1989; Cahill 2007).

While action research and participatory action research are suitable methodologies for a CCA approach, they are by no means exclusive options.

Ethnography, semistructured interviews, and focus groups are also suitable research methods as long as a level of reflexivity is incorporated, as we will explain. Researchers can select a combination of mixed methods and apply the principle of triangulation. The CCA can play an important role in interpreting findings and insights and provide guidance for data analysis and reflection on findings. Researchers also need to be sensitive to, and devise research methods that examine, power dynamics in projects. One way to do this is to ask respondents to draw diagrams of the structure of their projects, mapping stakeholders and drawing arrows to indicate flows of money, power, and influence.

An important recommendation derived from CCA is to incorporate reflexivity in the research and practice of open development. By reflexivity, we mean that the evaluation process is embedded in technical work and that the technical and evaluative work should inform each other. For the evaluation principles to be useful, a higher degree of reflexivity, which allows and requires participants not only to think about the best solution to a particular problem but also to question the framing and interpretation of the problem, is required. This will necessitate that individuals and organizations realize that the issues under evaluation are context dependent and need the specific attention of individuals with local knowledge and understanding.

Reflexivity furthermore requires explicit consideration of the role of the researcher within the research project. This research-related reflexivity needs to be embedded in the methodology, which means that the role that the researcher has in defining the research, selecting data sources, using data analysis mechanisms, and writing up findings should be open for questioning to both the researcher and other stakeholders in the research.

Identifying the issues using evaluation principles needs to be complemented by an identification of solutions. A difficult but important aspect is to be open about the description of the open application and its roles. Technologists may believe that they are the guardians of technology and therefore are in charge of providing definitions and descriptions. Similarly, donors might prescribe openness as a solution to a developmental problem and impose the terms under which an openness project is carried out. However, it is possible for there to be competing narratives that give rise to different views and capture the phenomenon as well or better. It is therefore important to realize that broad engagement of stakeholders is not only a democratic requirement but also a means to ensure that openness and its social consequences are understood appropriately.

Engaging with multiple stakeholders in research projects certainly raises significant challenges, especially in an academic climate that favors rapid research output. It requires long-term cultivation of relationships with people and institutions to support authentic participatory research. The CCA does not offer direct solutions to these challenges, except to point to the critical agency of researchers to reflect on their own roles in reproducing or shaping instrumental academic criteria and practices and to explore the possibility of resistance. While digitally enabled openness may afford researchers easy access to data and hence speedy output, the CCA would remind the researchers to consider which voices and narratives are absent or excluded from the digital arena and make efforts to engage with multiple stakeholders through digital or nondigital methods.

This short introduction should provide sufficient pointers for developing specific research approaches that use the CCA to understand and evaluate open development projects. The CCA does not offer an algorithm for research and does not constitute a methodology in its own right. Rather, it provides a theoretically rich starting point that can inspire research into how open development affects human freedoms and can motivate scholars to engage with this question with a view toward ensuring that such projects actually promote human development.

References

Ahari, Saeid S., Shahram Habibzadeh, Moharram Yousefi, Firouz Amani, and Reza Abdi. 2012. "Community Based Needs Assessment in an Urban Area; a Participatory Action Research Project." *BMC Public Health* 12 (1): 1–8. https://bmcpublichealth.biomedcentral.com/articles/10.1186/1471-2458-12-161.

Alkire, Sabina. 2005. "Why the Capabilities Approach?" *Journal of Human Development and Capabilities* 6 (1): 115–135.

Al-Taie, Mohammed Z., and Seifedine Kadry. 2013. "E-Government: Latest Trend and Future Perspective; The Iraq Case." *European Journal of Scientific Research* 99 (2): 307–323.

Argyris, Chris, and Donald A. Schön. 1989. "Participatory Action Research and Action Science Compared: A Commentary." *American Behavioral Scientist* 32 (5): 612–623.

Avgerou, Chrisanthi. 2010. "Discourses on ICT and Development." *Information Technologies & International Development* 6 (3): 1–18.

Avison, David, Richard Baskerville, and Michael Myers. 2001. "Controlling Action Research Projects." *Information Technology & People* 14 (1): 28–45.

Baguma, Johnstone. 2014. "Citizens' Advocacy for Public Accountability & Democratic Engagement through ICT Convergence in Eastern Africa." In *CeDEM14: Conference for E-Democracy and Open Government*, 449–462. Krems: Edition Donau-Universität Krems.

Baskerville, Richard L., and A. Trevor Wood-Harper. 1998. "Diversity in Information Systems Action Research Methods." *European Journal of Information Systems* 7 (2): 90–107.

Bentley, Caitlin M., and Arul Chib. 2016. "The Impact of Open Development Initiatives in Lower- and Middle Income Countries: A Review of the Literature." *Electronic Journal of Information Systems in Developing Countries* 74 (1): 1–20.

Boyd, Danah, and Kate Crawford. 2012. "Critical Questions for Big Data: Provocations for a Cultural, Technological, and Scholarly Phenomenon." *Information, Communication & Society* 15 (5): 662–679.

Brey, Philip. 2008. "The Technological Construction of Social Power." *Social Epistemology: Journal of Knowledge, Culture and Policy* 22 (1): 71–95.

Buskens, Ineke. 2011. "The Importance of Intent: Reflecting on Open Development for Women's Empowerment." *Information Technologies & International Development* 7 (1): 71–76. http://appropriatingtechnology.org/sites/default/files/Buskens%20The%20Importance%20of%20Intent_Reflecting%20on%20Open%20Development%20for%20Womens%20Empowerment%202011.pdf.pdf.

Cahill, Caitlin. 2007. "Including Excluded Perspectives in Participatory Action Research." *Design Studies* 28 (3): 325–340.

Chan, Leslie, Angela Okune, and Nanjira Sambuli. 2015. "What Is Open and Collaborative Science and What Roles Could It Play in Development?" In *Open Science, Open Issues*, edited by Sarita Albagli, Maria L. Maciel, and Alexandre H. Abdo, 87–112. Translated by Maria C. Matos Nogueira and Sandra C. Possas. Brasilia: Instituto Brasileiro de Informação em Ciência e Tecnologia; Rio de Janeiro: Universidade Federal do Estado do Rio de Janeiro. https://tspace.library.utoronto.ca/bitstream/1807/69838/1/What%20is%20Open%20and%20Collaborative%20Science%20book%20chapter.pdf.

Cilliers, Liezel, and Stephen Flowerday. 2014. "Information Privacy Concerns in a Participatory Crowdsourcing Smart City Project." *Journal of Internet Technology and Secured Transactions* 3 (3): 280–287.

Corrêa, Andreiwid S., Pedro L. P. Corrêa, Daniel L. Silva, and Flávio S. Corrêa Da Silva. 2014. "Really Opened Government Data: A Collaborative Transparency at Sight." In *2014 IEEE International Congress on Big Data*, Anchorage, Alaska, June 27–July 2, 2014, 806–807. Washington, DC: IEEE Computer Society.

Crawford, Kate, and Jason Schultz. 2014. "Big Data and Due Process: Toward a Framework to Redress Predictive Privacy Harms." *Boston College Law Review* 55 (1): 93–128.

Crocker, David A. 2008. *Ethics of Global Development: Agency, Capability, and Deliberative Democracy*. Cambridge: Cambridge University Press.

Crocker, David A., and Ingrid Robeyns. 2009. "Capability and Agency." In *Amartya Sen*, edited by Christopher W. Morris, 60–90. Cambridge: Cambridge University Press.

Cyranek, Günther. 2014. "Open Development in Latin America: The Participative Way for Implementing Knowledge Societies." In *Überwiegend Neuland: Positionsbestimmungen der Wissenschaft zur Gestaltung der Informationsgesellschaft*, edited by Dieter Klumpp, Klaus Lenk, and Günter Koch, 143–163. Berlin: Edition Sigma. http://www.instkomm.de/files/cyranek_opendevelopment.pdf.

Davies, Tim, and Fernando Perini. 2016. "Researching the Emerging Impacts of Open Data: Revisiting the ODDC Conceptual Framework." *Journal of Community Informatics* 12 (2): 148–178.

European Commission. 2013. "Directive 2013/37/EU of the European Parliament and of the Council of 26 June 2013 Amending Directive 2003/98/EC on the Re-use of Public Sector Information." *Official Journal of the European Union 56: L 175*. https://eur-lex.europa.eu/legal-content/EN/TXT/PDF/?uri=CELEX:32013L0037&from=EN.

Feenberg, Andrew. 1999. *Questioning Technology*. New York: Routledge.

Flor, Alexander. 2015. "Constructing Theories of Change for Information Society Impact Research." In *Impact of Information Society Research in the Global South*, edited by Arul Chib, Julian May, and Roxana Barrantes, 45–62. Singapore: Springer.

Gagliardone, Iginio. 2014. "'A Country in Order': Technopolitics, Nation Building, and the Development of ICT in Ethiopia." *Information Technologies & International Development* 10 (1): 3–19.

Graham, Mark, and Bernie Hogan. 2014. *Uneven Openness: Barriers to MENA Representation on Wikipedia*. Oxford: Oxford Internet Institute Report. https://ssrn.com/abstract=2430912.

Graham, Mark, Bernie Hogan, Ralph K. Straumann, and Ahmed Medhat. 2014. "Uneven Geographies of User-Generated Information: Patterns of Increasing Informational Poverty." *Annals of the Association of American Geographers* 104 (4): 746–764.

Gregson, Jon, J. M. Brownlee, Rachel Playforth, and Nason Bimbe. 2015. "The Future of Knowledge Sharing in a Digital Age: Exploring Impacts and Policy Implications for Development." IDS Evidence Report 125. Brighton: Institute of Development Studies.

Gurstein, Michael B. 2011. "Open Data: Empowering the Empowered or Effective Data Use for Everyone?" *First Monday* 16 (2). http://journals.uic.edu/ojs/index.php/fm/article/view/3316.

Hirchheim, Rudy, Heinz. K. Klein, and Kalle Lyytinen. 1995. *Information Systems Development and Data Modeling: Conceptual and Philosophical Foundations*. Cambridge: Cambridge University Press.

Janssen, Marijn, Yannis Charalabidis, and Anneke Zuiderwijk. 2012. "Benefits, Adoption Barriers and Myths of Open Data and Open Government." *Information Systems Management* 29 (4): 258–268.

Johnson, Jeffrey Alan. 2014. "From Open Data to Information Justice." *Ethics of Information Technology* 16: 263–274.

Johnstone, Justine. 2007. "Technology as Empowerment: A Capability Approach to Computer Ethics." *Ethics and Information Technology* 9 (1): 73–87.

Klein, Heinz K. 2009. "Critical Social IS Research Today: A Reflection of Past Accomplishments and Current Challenges." In *Critical Management Perspectives on Information Systems*, edited by Caròle Brooke, 249–272. Oxford: Butterworth-Heinemann.

Kleine, Dorothea. 2010. "ICT4WHAT?—Using the Choice Framework to Operationalise the Capability Approach to Development." *Journal of International Development* 22 (5): 674–692.

Kleine, Dorothea, David Hollow, and Sammia Poveda. 2014. *Children, ICT and Development: Capturing the Potential, Meeting the Challenges*. Florence: UNICEF Office of Research.

Linders, Dennis. 2012. "From E-government to We-government: Defining a Typology for Citizen Coproduction in the Age of Social Media." *Government Information Quarterly* 29 (4): 446–454.

Linders, Dennis. 2013. "Towards Open Development: Leveraging Open Data to Improve the Planning and Coordination of International Aid." *Government Information Quarterly* 30 (4): 426–434.

Martin, Michael, Claus Stadler, Philipp Frischmuth, and Jens Lehmann. 2014. "Increasing the Financial Transparency of European Commission Project Funding." *Semantic Web* 5 (2): 157–164.

McLennan, Sharon J. 2016. "Techno-optimism or Information Imperialism: Paradoxes in Online Networking, Social Media and Development." *Information Technology for Development* 22 (3): 380–399.

Miller, Jeff. 2009. "Open Education Resources: Providing Incentives for Commons-Based Production Activities in Higher-Education." *Access to Knowledge: A Course Journal* 1 (2): 1–12.

Mumford, Enid. 2001. "Advice for an Action Researcher." *Information Technology & People* 14 (1): 12–27.

Njihia, James Muranga, and Yasmin Merali. 2013. "The Broader Context for ICT4D Projects: A Morphogenetic Analysis." *MIS Quarterly* 37 (3): 881–905.

Oosterlaken, Ilse. 2011. "Inserting Technology in the Relational Ontology of Sen's Capability Approach." *Journal of Human Development and Capabilities* 12 (3): 425–432.

Panek, Jiri. 2015. "How Participatory Mapping Can Drive Community Empowerment—a Case Study of Koffiekraal, South Africa." *South African Geographical Journal* 97 (1): 18–30.

Peixoto, Tiago, and Jonathan Fox. 2016. "When Does ICT-Enabled Citizen Voice Lead to Government Responsiveness?" WDR 2016 Background Paper. Washington, DC: World Bank Group.

Peled, Alan. 2013. "Re-designing Open Data 2.0." *Journal of E-Democracy and Open Government* 5 (2): 187–199.

Phang, Chee Wei, Atreyi Kankanhalli, and Lihua Huang. 2014. "Drivers of Quantity and Quality of Participation in Online Policy Deliberation Forums." *Journal of Management Information Systems* 31 (3): 172–212.

Rawls, John. 1971. *A Theory of Justice*. Cambridge, MA: Harvard Business Press.

Reilly, Katherine M. A. 2014. "Open Data, Knowledge Management, and Development: New Challenges to Cognitive Justice." In *Open Development: Networked Innovations in International Development*, edited by Matthew L. Smith and Katherine M. A. Reilly, 297–326. Cambridge, MA: MIT Press; Ottawa: IDRC. https://idl-bnc-idrc.dspacedirect.org/bitstream/handle/10625/52348/IDL-52348.pdf?sequence=1&isAllowed=y.

Robeyns, Ingrid. 2005. "The Capability Approach: A Theoretical Survey." *Journal of Human Development* 6 (1): 93–117.

Robeyns, Ingrid. 2006. "The Capability Approach in Practice." *Journal of Political Philosophy* 14 (3): 351–376.

Sandoval-Almazán, Rodrigo. 2015. "Open Government and Transparency: Building a Conceptual Framework." *Convergencia Revista de Ciencias Sociales* 22 (68): 203–227.

Sen, Amartya. 1984. *Resources, Values and Development.* Oxford: Basil Blackwell.

Sen, Amartya. 1987. *The Standard of Living*. Cambridge: Cambridge University Press.

Sen, Amartya. 1993. "Capability and Well-being." In *The Quality of Life*, edited by Martha Nussbaum and Amartya Sen, 30–52. Oxford: Clarendon Press.

Sen, Amartya K. 1999. "Democracy as a Universal Value." *Journal of Democracy* 10 (3): 3–17. https://www.journalofdemocracy.org/article/democracy-universal-value.

Sen, Mourjo, Anuvabh Dutt, Shalabh Agarwal, and Asoke Nath. 2013. "Issues of Privacy and Security in the Role of Software in Smart Cities." In *2013 International Conference on Communication Systems and Network Technologies*, Gwalior, India, April 6–8, 2013, 518–523. Washington, DC: IEEE Computer Society.

Smith, Matthew, and Laurent Elder. 2010. "Open ICT Ecosystems Transforming the Developing World." *Information Technologies & International Development* 6 (1): 65–71.

Smith, Matthew, Nathan J. Engler, Gideon Christian, Kathleen Diga, Ahmed Rashid, and Kathleen Flynn-Dapaah. 2008. *Open ICT4D*. Ottawa: IDRC.

Smith, Matthew L., and Katherine M. A. Reilly, eds. 2013. *Open Development: Networked Innovations in International Development*. Cambridge, MA: MIT Press; Ottawa: IDRC. https://idl-bnc-idrc.dspacedirect.org/bitstream/handle/10625/52348/IDL-52348.pdf?sequence=1&isAllowed=y.

Smith, Matthew L., and Ruhiya K. Seward. 2017. "Openness as Social Praxis." *First Monday* 22 (4). https://firstmonday.org/ojs/index.php/fm/article/view/7073.

Stahl, Bernd C. 2008. "The Ethical Nature of Critical Research in Information Systems." *Information Systems Journal* 18 (2): 137–163.

Stahl, Bernd C. 2011. "Critical Social Information Systems Research." In *The Oxford Handbook of Management Information Systems: Critical Perspectives and New Directions*, edited by Robert D. Galliers and Wendy L. Currie, 199–228. Oxford: Oxford University Press.

Susha, Iryna, Anneke Zuiderwijk, Marijn Janssen, and Åke Grönlund. 2014. "Benchmarks for Evaluating the Progress of Open Data Adoption: Usage, Limitations, and Lessons Learned." *Social Science Computer Review* 33 (5): 613–630.

Terrell, Josh, Andrew Kofink, Justin Middleton, Clarissa Rainear, Emerson Murphy-Hill, Chris Parnin, and Jon Stallings. 2017. "Gender Differences and Bias in Open Source: Pull Request Acceptance of Women versus Men." *Peer J Computer Science* 3: e111.

Thurston, Anne C. 2012. "Trustworthy Records and Open Data." *Journal of Community Informatics* 8 (2). http://ci-journal.net/index.php/ciej/article/view/951.

Trapnell, Stephanie E. 2014. "Right to Information: Case Studies on Implementation, Right to Information." Right to Information Working Paper Series. Washington, DC: World Bank Group.

Wen, Sophia M.-L., and Tze-Chang Liu. 2016. "Reconsidering Teachers' Habits and Experiences of Ubiquitous Learning to Open Knowledge." *Computers in Human Behaviour* 55 pt B: 1194–1200.

Zheng, Yingqin. 2009. "Different Spaces for E-development: What Can We Learn from the Capability Approach?" *Information Technology for Development* 15 (2): 66–82.

Zheng, Yingqin, and Bernd C. Stahl. 2012. "Evaluating Emerging ICTs: A Critical Capability Approach of Technology." In *The Capability Approach, Technology and Design*, edited by Ilse Oosterlaken and Jeroen van den Hoven, 57–76. Dordrecht: Springer Netherlands.

Zheng, Yingqin, and Geoff Walsham. 2008. "Inequality of What? Social Exclusion in the E-society as Capability Deprivation." *Information Technology & People* 21 (3): 222–243.

Zuboff, Shoshana. 2015. "Big Other: Surveillance Capitalism and the Prospects of an Information Civilization." *Journal of Information Technology* 30 (1): 75–89. https://doi.org/10.1057/jit.2015.5.

10 Open Institutions and Their "Relevant Publics": A Democratic Alternative to Neoliberal Openness

Parminder Jeet Singh, Anita Gurumurthy, and Nandini Chami

The Field of Open Development

Information and communications technologies (ICTs) are transforming our societies in fundamental ways. Early on, ICTs were largely regarded as useful *tools* to better achieve various social and economic objectives, an approach that underpinned the field of ICTs for development (ICTD) (Rosenberger 2014). It has increasingly become evident, however, that the social impact of ICTs is deeply structural, and theorists have sought more robust concepts and theories to capture the role of ICTs in development.

Open development is one such attempt. ICTs indeed deconstrain information flows and social relationships and thus can be considered as promoting openness, possibly leading to positive results. In an earlier paper (Singh and Gurumurthy 2013), we had broadly defined openness to mean decreased constraints on social interactions. *Openness* is now well understood in relation to certain sectors, such as open government, open knowledge, and open technology. It has also been applied to some other fields, such as education, science, and health. In each of these areas, there are a set of benchmarks, some more accepted than others, to suggest whether and why a practice belongs to open X, X being government data, knowledge, educational resources, and other materials.[1] Such benchmarks are contextual to the area or sector and do not necessarily apply across sectors. One way to understand open development is to consider it a field that encompasses the aggregate of all the *open X* forms, with a broad family resemblance (Davies 2012).

But this begs the question, why do we need a distinct term called *open development*? The idea behind open development, as argued by Smith and Reilly (2013a, 4), is to harness "the increased penetration of information and communications technologies to create new organizational forms that improve the lives of people." Thus, it can be said that "networked organizational forms" provide development practice with a new paradigm to effect change. The theoretical field of open development, however, reflects a tension between the normative and the positivist character of openness. Smith and Reilly (2013b, 15), for instance, observe that "new open networked models ... will not necessarily lead to social good." Other scholars, such as Buskens (2013, 341), consider open development as "enabling and enhancing equity, sharing, and connectedness" and "as a critique on the existing utilitarian, growth-driven, econocentric mainstream worldview."

Our starting point is that development is certainly a normative discipline, and open development also needs to be seen as such. The visible impacts of ICTs are often powerful and spectacular. It is hence vital to distinguish between what may simply be far-reaching changes arising out of the new organizational possibilities afforded by ICTs and those that endure as equitably beneficial to people. Emphasizing the end goals of open development is particularly important given recent technological advancements. The last couple of years have highlighted that big data, which demands different theoretical treatment than information, and "digital intelligence" may be the real game changers (Singh 2017, 1). As a social nervous system running across all sectors, big data and artificial intelligence (AI) are likely to fundamentally change the way our social institutions are organized (Singh 2017).

Benkler (2006, 32) considered "radical decentralization of intelligence in our communications network" to be the key contemporary transformative force. With big data and AI-based processes, we are witnessing an opposing trend—a movement toward the radical *centralization* of intelligence across our social systems. This has direct implications for how the benefits of ICTs will be distributed.

Our approach takes a normative view of open development, laying out the building blocks of how the concept connects to a better distribution of power in society and its various institutions. The open Internet, as a global communication system with no central control or administration, and its unregulated institutional ecosystem based on *free*, although actually nontransparent and unequal, contracts among private Internet service providers (ISPs) along with users, have over time seen a neoliberal takeover by big corporations (McChesney 2013). In fact, the Internet's openness itself is employed as a mantra for a new social design, which is what gives the term *openness* its newfound currency and vitality.

Google's smart cities company, Sidewalk Labs, is conceptualizing a city that is developed from the "Internet up," where the organizational logic and efficiencies of the Internet will ab initio inform a city's various systems (Doctoroff 2016). Technology evangelists speak breathlessly about smart solutions to practically every social problem, which are based on new, disruptive digital business models. The successful businesses get cannibalized by big corporations, which are orchestrating the economic reorganization of whole sectors through platform-based digital ecosystems. Jack Ma, the head of China's e-commerce giant Alibaba Group Holding Limited (2016), has proposed a new global digital trade platform involving virtual free trade zones.

Our engagement with the normative aspects of open development is a response to such digitally enabled social reorganization, wherein *openness* is appropriated as the key value and premise to further a neoliberal private ordering (Morozov 2013).[2] We seek to posit how the promise of ICTs and their *openness* can be captured for a new age democratic public order. Our concept of democracy proceeds from the standpoint of ordinary citizens. The new democratic possibilities from ICT-induced openness must privilege the participation and empowerment of marginalized citizens, countering what Gaventa (2006, 21) describes as "neoliberal or liberal representative understandings" that "often remain hegemonic."

Openness and *open development* must be placed in relation to a larger institutional framework, going beyond their application to specific artifacts or processes. ICT-based affordances create the possibility for new institutional designs wherein all stakeholders or communities relevant to an institution are more closely and continually able to deliberate and influence the institution and its functions. We refer to such institutional redesign as ICT-induced *openness* and the resulting institutions as *open institutions*. We consider such an institutional transformation process as a generalized socialization of the idea of participatory democracy, as it gets applied not only to major political organizations but to all social organizations.

The first section of the chapter provides a definition of *open development* as consisting of a movement toward *open institutions*. This definition places the customary constituents of *openness*—transparency, participation, and collaboration—in a situated institutional setting, as a set of social relationships among specific social actors. The next section focuses on the social actors who enjoy an enhanced role in the work and positive outcomes of an institution that is opening up. These actors are defined as the "relevant public" of the institution. The following section shows how such an institutional definition, focused on the crucial element of *public interest*, addresses the key problem of *open washing*, whereby practices that do not actually benefit the wider community nonetheless get promoted as *open* processes or initiatives. The next section explores how *public interest* is determined, and its fulfillment judged, in the specific context of an institution whose openness is being examined. A discussion of how the proposed paradigm of *open institutions* and *open development* can form the basis of a new democratic reordering of society, rather than a neoliberal ordering, which is the dominant digital macrotrend, follows. The chapter concludes by proposing that such democratic *opening* of institutions is the most appropriate way to address the problem of institutional capture that ails our democratic political systems, often providing the justification for their neoliberal, corporatist replacement.

Open Development as Open Institutions

We propose an institutional definition of *open development* as the use of ICTs for institutional redesign to bring about structural changes that enhance transparency or information sharing, participation, and/or collaboration, in a manner that is primarily motivated by, and contributes to, the public interest. This definition foregrounds institutional change, defined as change in entire classes of organizations, causing deep shifts in the ideas that govern institutions and consequently the rules and practices associated with such ideas (Halal 2008). Open institutions consist of ideas, rules, and practices of openness that change a whole class of organizations.[3] While our definition focuses on institutions and deep change, we also refer in our discussion to organizations as particular instantiations and sites of such institutional change. One can speak about organizational or institutional change, depending on how deep and far-reaching a change is.

To provide an example illustrating the difference between anchoring a change theory at an organizational and at an institutional level, let us consider the institution of school versus a particular school as an organization. On the one hand, we can speak about changes in a school, such as closer teacher-student interaction or the ease with which students can opt in or out of specific subjects. Such organizational changes can lead a particular school to achieve different educational outcomes than similar schools. On the other hand, considering what is happening with the advent of digital possibilities, completely new paradigms of personalized learning are possible. Students can self-determine their pace of learning, the role of the teacher is also transformed, and methods of both teaching and learning change dramatically. As the new paradigm is absorbed across society, a shift affecting the very institution of the school is evident. Analyses of institutional and organizational changes are not mutually exclusive but take place along a continuum.

Returning to our definition of open development, in theory, greater transparency through information sharing, participation, and/or collaboration tends to cause a better distribution of power in favor of the community of stakeholders that an institution or organization caters to. This enhances the achievement of public interest inherent in the social function of that institution or organization and reduces its capture by insiders or holders of powerful roles.

Our definition of *open development* contains six elements. Open development is about (1) institutions; (2) the context of the social impact of ICTs; (3) deliberate design; (4) the wider structural changes effected; (5) how such changes increase transparency, participation, and/or collaboration; and (6) ensuring that the changes are motivated by, and contribute to, the public interest.

Development is generally recognized as consisting of sustained institutional change (Chang 2007). A theory of open development must therefore be anchored in relation to institutions. As an open education practitioner observed, "People make things possible. Institutions make them last" (Caulfield 2016, n.p.). Moreover, it is at the institutional level that the nature of different interests and power relationships, which are basic to understanding development, can be appropriately observed and analyzed.

The concept of open development arose as an attempt to understand and analyze ICT-induced social changes. The second element of the definition relating to the context of the social impact of ICTs is therefore a priori in charting a field of interest.

The third element draws on development as a set of deliberate actions that are generally collective, whether the deliberateness is explicit (as in public spending) or implicit (as in the social contract). It should be noted that in espousing this position this chapter departs from the methodological individualism of mainstream openness theorizations, which is its key difference with other approaches in this area. We believe that social facts are best illuminated through the application of multiple lenses and perspectives. As an illustration, it is not the discrete daily actions of thousands of Wikipedia contributors that by themselves comprise open development but instead the coming into existence of rules and conditions for such actions, which obviously takes place at a higher institutional level. As the contributor community evolves and the emergent and explicit rules and norms are integrated, the institutional system acquires some degree of collective deliberateness. The *deliberate design* element of the definition is important because development normally refers to a process of induced (institutional) change and not merely performance.

Fourth, open development changes must not be one-off but should be structural, producing conditions for stable, positive outcomes. These changes should result in sustained enhancement in one or more of the three generic social processes that are commonly enabled by ICTs in institutions and organizations: transparency, which includes, for the purposes of this discussion, access to informational resources; participation; and/or collaboration. A focus on these processes has been the tradition in the field of *openness* and *open development* (Cyranek 2014; Harvey 2011; Smith and Elder 2010). Regarding how and why institutional and structural changes affect transparency, participation, or collaboration, we take a situated, power-relational view of these processes. To consider transparency (or information sharing), participation, or collaboration as standalone actions or processes is not very meaningful and does not provide much analytical or theoretical value in a social change or development context. These processes must hence be seen as being embedded in social relationships with implications for social power.

Both transparency and participation clearly connote a set of inside and outside actors. Collaboration may not inherently reflect such an inside-outside quality and may be taken to be distributed across a large number of actors. However, given that collaboration is rarely an evenly distributed process, it is normally possible to identify a core, organizing space and an outside, community space in relative terms, even if there is a certain continuum between the two. Openness, as enhancement in one or more of these three processes, therefore refers to a change in the nature of the corresponding social relationships. Such a change almost always entails a power shift between two kinds of actors: organizational actors and what can be called the relevant public.

Regarding organizational transparency, information is often considered a nonrivalrous resource (sharing it costs nothing or a negligible amount) to the holder of information, who may even benefit from such sharing. Information's real value, however, lies mostly in its application to material contexts. This value is generally rivalrous. Someone may lose as someone else gains. Organizational transparency comes mostly at the cost of the interests of organizational inside actors even as it benefits outside actors. This is why, as the open data literature suggests, we find that government initiatives often put out information that is politically insignificant for citizens (Cañares 2014; Michener 2015). Transparency, then, is not about the extent of organizational information shared but instead about the public interest intention that is involved, which can be judged by the resultant power shifts, if any, toward outside actors. Hence, providing access to organizational information can be considered transparency only if seen in the context of a relationship between an organization and its outside actors and when it leads to a shift in power in favor of the latter. Such a shift would mean that the public interests, or the outside actors, are now better represented in organizational operations and outcomes than before.

Similarly, allowing outside actors to share their views on organizational matters is not by itself increased participation. It could merely be providing "voice without agency" (Singh and Gurumurthy 2013). For it to constitute participation, any such action must change the power relationship between actors inside the organization and the outside actors. Likewise, collaboration, in the openness context, implicitly favors distributed power-sharing structures. Collaboration involves a large number of distributed actors creating value together and is also expected to lead to a better distribution of such value. Crowdsourcing free or underpaid labor for corporate gain cannot be considered collaboration. Collaboration is not only a social relationship with power implications. True collaboration implies a certain pattern of power distribution that does not discriminate unduly against actors who may not be centrally involved with organizing the collaboration.

Thus, it is apparent that the terms transparency, participation, and collaboration are only meaningful when they involve a shift of power from a set of inside (organizational) actors to those outside (community). For an institution undergoing openness, the important category of *outside actors* whose power must be enhanced, if these processes are to be effective, is the *relevant public.*

One of the most significant elements of the definition concerns the introduction of the concept of *public interest.* Public interest represents the widest articulation of social good or positive social objectives. We have avoided getting into specific definitions of the social objective(s) involved, which could be social development, economic growth, and building people's capabilities, among others. In a context where *openness* is increasingly being co-opted as a core constituent of neoliberal reordering of society (Betancourt 2016; Foster and McChesney 2011), a clear distinction between *public* and *private* interest is important. Open development must not just contribute to the *public interest* but also be motivated *primarily* by an intention toward public interest. Activity motivated by private interest may masquerade as *open development* and may contribute to public interest in the short term, but the sustainability of that contribution cannot be guaranteed. This has caused the pernicious phenomenon of *open washing,* which will be discussed later in this chapter.

Invocation of *public interest* in our definition also brings a focus to the actual recipient of the benefits of *openness*. At the highest generalization, it could just be the larger public, as understood in political and media theories, but one of the key ideas that our definition of *open development* contributes is identifying a specific public for a given institution or organization, as the group that *receives* and benefits from the latter's openness.

Relevant Publics

If *open development* is about developing new ICT-based processes for greater transparency, participation, and collaboration *with respect to* a specific *relevant public,* it becomes important to be able to identify such a public. The traditional notion of a general public, which is employed by political and media theory, is not appropriate in this context because it is not possible for institutions to develop adequate ICT-enabled relationships with everyone, nor would it be efficient to try to do so. In fact, it is because of the nebulous nature of the undefined *public* that public interest organizations may escape accountability to their constituencies. This can result in institutional capture, a concept that is discussed in detail later.

At the outset, it should be clarified that in proposing our concept of *relevant publics,* we use the term *public* in the sense of a political agent rather than in the discursive sense of a public sphere. We build on John Dewey's (1927) concept of *public* as a category that arises in response to a collective problem. Insofar as any public institution is supposed to be constituted in relation to some collective or public problem (and the problem, in positive terms, could be how to utilize an opportunity), it makes sense to speak of relevant publics for public interest organizations. The public sphere also has a structural impact on political actions, and new theories of networked public spheres should have relevance to the concept of *relevant publics*. Such an exploration is not attempted in this chapter, but it remains an important area for future study.

Most definitions of openness aspire to a universal reach of openness—of transparency, participation, and/or collaboration. While morally justifiable, there are practical and theoretical problems with addressing a universal set of outside actors. Effective ICT-enabled structural relationships between institutions and their relevant publics can only be developed if the latter are identified with a sufficient degree of clarity. The ICT-based means for such relationships need to be contextual and made available proactively to the specific relevant public. Many critiques of open data practices illustrate this point; it is not enough to simply put data out there for *anyone* to access and use.[4] To be meaningful, open data practices must be oriented toward the actual needs of their specific *relevant publics*.

Open data should accordingly be structured and presented in an appropriate manner, and adequate means should be ensured for its purposeful use by the group(s) that it is intended for. For instance, Reilly and Alperin (chapter 2, this volume) stress that actors engage in regimes of open data stewardship that create different types of social value. For data to be considered really open in an institutional sense, the actors, stewardship regimes, and value created for specific publics must be understood. One way to determine the relevant public could be through self-identification: whoever shows interest in relating to a public interest organization constitutes its relevant public. Even so, to design effective *openness* relationships, the concerned public interest organization would need to make an assessment of what its relevant public actually is. In many cases, self-identification may not be the appropriate criterion at all.

Where rivalrous resources are involved, it may be necessary to restrict them for use by certain publics, excluding others. This would require rules-based delimitation of the relevant public. Elinor Ostrom considers laying clear group boundaries (Walljasper 2011) the first principle for the management of commons, a concept that is related to openness.

Legitimate exclusions can apply even for access to information, normally regarded as nonrivalrous, because, as stated earlier, the real value of information is mostly in its application to material (and rivalrous) contexts. A local farmers' collective may share agricultural information among its members but justifiably exclude commercial agriculture companies from accessing it. Participation in decision-making and collaboration, for production and distribution, has even clearer justifications for the marked-out boundaries of relevant publics. Furthermore, it is important to note that not all kinds of collections and aggregations of interest can be called publics. Publics must uphold some generally accepted definition of public interest, such as promoting farmers' livelihoods. But a business cartel would not qualify, although many kinds of business associations will be considered as being in the public interest. Publics must also be sufficiently inclusive. Arbitrary exclusions would not meet the public interest criterion.

It is often just the nature and objective of a set of activities, projects, or organizations to cater to a clearly identified group, even while working for the public interest. An education system that freely shares educational resources only among those registered as students in a country may still qualify as an instance of open development. Principles of open development would, in turn, encourage providing students some degree of say regarding which resources are made available and how. A community development project can have as its relevant public a small community or a section of it, say the adolescent girls of that community. Free sharing of informational resources only within such a relatively small group may not make for a huge openness claim, but employing ICTs to include the group in decision-making and developing resources together renders it an open development project.

Definitions of openness or open development that are centered on process, action, or artifacts, demanding universal access, participation, and collaboration, are unable to adequately account for cases where there is a *justifiably* limited or circumscribed reach of openness. Identifying a relevant public through contextualized rules and relationships helps researchers and practitioners avoid glossing over who exactly openness intends to serve.

The Problem of *Open Washing*

Analyzing motivation for public interest is more important for recognizing *open development* than understanding outcomes alone. Determining the motivation for public interest is necessary to counter the potential for corporate and other powerful organizations to co-opt openness to pursue their own interests. Open washing refers to instances when institutions claim to practice openness but engage their *relevant publics* only in a limited and self-serving manner—for instance, when companies offer "free" data but require that users employ proprietary platforms to actually make use of the data, as is in the case of Facebook's Free Basics. Open development requires the means to identify underlying motivations instead of relying only on the immediate outcomes of organizational practice.

Shifts in relative power between inside and outside actors (the *relevant public*) related to an institution or organization are better assessed from the intention of a public interest. However, organizations are often judged largely based on their performance related to a series of outcomes. Yet an outcome may be incidental and secondary to an actor's primary motive. Intention, on the other hand, is normative. Although more difficult to assess than actual outcomes, it represents a lasting quality of an organization. Structural changes, which we are interested in here, may or may not immediately cause a positive outcome. Likewise, an immediate positive outcome may not represent a structural change and hence a sustained result. It is even possible that some immediate outcomes that appear positive may actually be related to structural changes that are harmful in the mid- to long term. The manner in which major digital companies, formerly celebrated as exemplars of digital virtues, are currently facing strong public backlash over issues such as privacy, net neutrality, anticompetitive behavior, and tax avoidance testifies to this fact.

The real intentions of an organization in multiactor conditions may not be clearly evident as long as a win-win situation prevails. However, such intentions and the relative power between actors are revealed when conflict between the interests of different actors arises. Conflicts of interest are normal in any context where private interests are being pursued, and they are bound to surface sooner or later.

It is especially important to make a distinction between public and private interests in times of rapid technological change, as is being witnessed currently. When new technologies provide everyone many new opportunities that were previously unknown, it is typically viewed as a win-win situation between technology providers and users. In this context, the private interests of technology providers may be obscured owing to the power differentials between such providers and technology users. When rapid technological change leads to structural shifts and new social designs is precisely when distinctions between public and private interests are most important to make in order to form appropriate norms to guide social action in new situations. Unfortunately, concepts and theories of openness have not been very successful in meeting this imperative. In fact, they have likely been guilty of adding to the confusion.

Today's digital corporations are the agencies behind large-scale innovation, production, and provision of digital goods and services. Users experience enhanced agency, especially in terms of freer information flows and communication, because of the digital possibilities made available by these corporations. Many key digital services are provided apparently free (as the personal data-related costs remain invisible), and in a monopolistic fashion, because of the winner-takes-all nature of the sector. This makes the corporations look like pro bono providers of public goods—further bolstering their false image of openness oriented toward the public interest. Prior slogans such as Google's "Don't be evil" and Facebook's "Digital Equality" present corporate activities as being motivated primarily by the public interest. Major digital corporations have co-opted the virtues and good image of openness to further their business interests, and they benefit from the cooperation of public interest actors by projecting win-win situations. Such acts have been referred to as *open washing* (Murillo 2017).

When Google publicly shares its maps and enables access to application programming interfaces (APIs), loose conceptions of *openness* allow Google and others to call it an act of openness, undertaken in the public interest. This is similar to the case of people's *participation* in shaping trending news topics on Facebook. Such actions and processes motivated by private interest that may secondarily contribute to public interest must be distinguished from those primarily motivated by public interest. Whether the nature of actions and outcomes is easily distinguishable or not, norms framing private or public interests are clearly different. The latter should therefore be the focus of an institutional inquiry. Such an inquiry is important because monopolistic digital corporations have illegitimately squatted on the field of openness, making digital openness a contentious subject among public interest and development actors.

Organizational intention is a matter of norms, providing a better measure of public interest than outcome. As discussed, such norms and intentions are most discernible in situations of conflict of interest. It is useful, for instance, to see how digital corporations react to proposals for public interest regulations in the digital sector in areas such as net neutrality, privacy, and tax avoidance.

Open-washing digital corporations not only derive good publicity from the endorsements made by public interest actors. They increasingly use these actors as a cover to entrench very exploitative power structures in emerging digitally mediated societies. The ostensibly *free and open* global digital infrastructure, which resists public interest regulation, has become an open mine for the most important resource in the digital society—big data.

The primary focus of practically every large digital business today is on building *digital intelligence* (Singh 2017) by invasively collecting personal data. Such intelligence provides rentier positions that can be exercised to control entire sectors globally. Seeking open data flows across borders is the typical *openness* pitch in this regard. Both privacy and trade justice activists have found this problematic.[5] Such new digital developments, where digital intelligence is the central transformational factor, are not separable from processes or actions that usually get studied under the openness rubric. These developments have complicated the field of open development. In times of such complex society-wide changes, any theory that uncritically focuses on a narrow set of promising processes and characteristics stands on weak ground. An institutional approach to open development attempts to address these shortcomings.

Contextual Public Interest for an Open Organization

We have defined open development as a shift by organizations toward sustained ICT-enabled sharing, participation, and/or collaboration, with the aim of furthering public interest. A central issue remains: how is public interest determined? This is in fact the central question with regard to the political organization of our societies.

It is possible to determine whether an organization is motivated by public interest by establishing the primacy of public interest in the institution's organizational design, inferred through the rules and sustained actions of the organization.[6] This constitutes a technical determination of *public interest motivation* through institutional analysis. It is enough to employ such an institutional analysis of public interest motivation to practically apply our definition of *open development*, both for evaluation and for new organizational designs. However, some subjective elements remain in such a determination.

Adam Smith claimed that people acting per their private interests form the "invisible hand" that best ensures public interest, at least for economic production and distribution. Neoliberals seek to extend this invisible hand to all social affairs. Philanthropists and civil society groups determine public interest on the basis of private knowledge or inclinations. They also employ experts frequently. Mark Zuckerberg may say that he runs Facebook in the public interest. If people point to the profit-making aspects of the business, it can be justified as necessary to sustain the public interest work of Facebook. It can therefore become difficult to ascertain public interest, not only from the nature of outcomes but also from intentions, when the profit motive can be presented merely as instrumental to the larger public interest orientation. To address this subjectivity problem, we would like to take our theory beyond technical determination by exploring the possibilities for democratic-political determination of public interest, directly and explicitly, by the *relevant publics* within the specific context of a given organization or institution engaged in *open development.*

Perhaps the only dependable measure of public interest is what the public considers public interest, although there are enough democracy skeptics who raise doubts as to whether the public best understands what public interest is or is capable of meaningfully expressing it (Lippmann 1927). The public conveys its interest in elections to constitute state power. Public interest is also conveyed, albeit much less definitively and effectively, through participatory democracy, which is mostly aimed at the state, and through the public sphere. Regarding state power, today there is great dissatisfaction with political processes, including elections. In any case, such an articulation of public interest, at the level of the nation, the state, or even locally, does not translate well into actual institutional design or the work of various public bodies, as well as public interest organizations outside the public sector. The open development approach is most concerned with these meso- and microlevels.[7] This is not to undermine the relevance of state-centric articulation of public interest, conveyed to these institutions through political, legal, bureaucratic, and regulatory processes that are important to society's political and social organization.

One of the most promising features of the network society is that entirely new opportunities have opened up for determining what public interest is, not only for the overall polity as traditionally done but also for the specific contexts of a given institution or organization. With public interest organizations developing close ICT-based structural relationships with the sections of the public most affected by its functions, a granular kind of participatory democracy at the meso- and microlevels is made possible. Such relationships pertain both to ICT-based transparency and to participation in decision-making, the precise nature of which will be contextual to the organization concerned.

In the current times of prevalent distrust of institutions across the globe, it is often cynically concluded that public interest is simply what the people controlling institutions consider or present to be public interest. This is called *institutional capture*. A university or a public health body, for instance, may selectively undertake some kinds of ICT-based structural changes and justify them on the basis of public interest; for example, by adopting ICTs to carry out some of its administrative tasks but not to make the functioning more transparent and inclusive. Some of these actions or nonactions may not be considered to be in the public interest by outsiders. Increasingly, the response to the institutional capture problem has been to subject more and more social institutions to private sector practices, which is also called the *neoliberalization* of society. New Public Management, for example, is one such trend (Vabø 2009).

The promising alternative to this problematic neoliberal so-called solution is for public interest institutions to employ ICT-enabled openness to develop deeper and more stable structural relationships with their relevant publics, enabling better contextual discovery and fulfillment of the public interest. Here, the public interest requirement for *openness* of an organization is determined not just by norms shaping its outcomes but also from adoption of democratic participation of the *relevant publics* in its decision-making processes. Such new relationships can improve the efficiency of institutions and organizations as well as contribute to making them more just, reordering them in ways that are democratic.

Such democratic *openness* is as relevant to the voluntary sector as to the public sector. The former works in the name of public interest but without any clear means of assessing what public interest is. When public interest institutions and organizations develop such direct and horizontal relationships with their relevant publics—through ICT-enabled transparency, participation, and collaboration—this constitutes *open development*.

Public Systems in the Network Age

Open institutional design allows closer interaction between public interest organizations and their relevant publics. This section first delineates how such a relationship can be materially structured around the elements of transparency, participation, and collaboration. An *open institution* must, by default, make information about itself available to the relevant public and, to the extent possible, to the whole public. Any nondisclosure must be clearly justified by predefined exceptions. These exceptions should be open to discussion and the influence of the relevant public through digital (and other) means. Similar participation must extend to all key organizational decision-making processes. Depending on the nature of public interest work, the relevant public should be provided with ICT-enabled and other collaborative avenues for developing, as well as utilizing, the organization's resources. Such a collaborative approach to developing public goods is especially relevant for digital public goods but not limited to them. Considerations of efficiency will need to be taken into account in all these new processes without allowing them to become a cover for insiders' vested interests.

In the pre-ICT/Internet age, because of transactional constraints, the default for large public interest institutions and organizations was set to nearly zero horizontal interaction with outside actors. This meant general nontransparency, nonparticipation, and noncollaboration, with very selective possibilities on an as needed basis for information sharing, participation, and collaboration. In open institutional design, with the cost and means of large-scale interaction across distance completely transformed, the default should be set in the reverse direction. There needs to be a clear and demonstrated need and specific reasons to close off information sharing, participation, and collaboration. Such an open by default criterion has already been applied to information-sharing practices by governments in the *open government* field.[8] This needs to be broadened across all public interest organizations, including the public interest aspects of private organizations,[9] and it should be taken beyond information sharing to include participation and collaboration.

Employing these criteria, it is possible to evaluate the extent to which public interest organizations have *opened up* or are pursuing *open development*. This approach also provides the benchmarks and design principles for new organizations. Practicing open development will mean that existing organizations undertake a design overhaul to meet these requirements to the maximum extent possible. New organizations must begin with such a template as the default design before contextual features are added. Despite what it seems, this is not such an extraordinarily drastic prescription; almost every business of a considerable size today is undergoing significant redesign to cater to the context of the digital network society. Public interest organizations cannot afford to lag behind in a context where a historically significant contest is under way between the public interest sector and the corporatist organization of society. The relative effectiveness of digital mutations in these two areas may turn out to be crucial to the outcome of this contest. While an organization's identified relevant public is immediately most significant to it, accountability to other publics, including the overall public, is important. Publics then become a graded and networked system, with some degree of hierarchies.

Earlier, we argued why the concept of public-ness is better than openness at capturing new development possibilities arising from ICT adoption (Singh and Gurumurthy 2013). We conceptualized the term *network public* as (Singh and Gurumurthy 2013, 188) "much more than the 'networked public sphere' described by Yochai Benkler and others. Network public covers a much wider public institutional ecology, consisting of various public and community institutions in their diverse functions. Basically, the network public represents the public segment or aspect of the network society, formed of its spaces, and its flows."

The public sector produces and provides public goods. In the network society, this function is best performed through a networked system: "A network public model will consist of networks of public authorities, development agencies, progressive techies, and the community in general, working together to build and sustain various digital and socio-technical artifacts and platforms that underpin our digital existence (software, social media, search engines, and so forth)" (Singh and Gurumurthy 2013, 188). This description is an example of producing digital public goods, but the concept applies to all kinds of public goods.

The *network public* was described as a network age innovation "at the boundaries between the state and the community" (Singh and Gurumurthy 2013, 188), and the concept of deepening democracy was proposed as a good starting point for it. A network public includes "creat[ing] an effective space for development dialogue and discourses" (Singh and Gurumurthy 2013, 189).

In the paper just mentioned, we had critiqued open development models and theories as being not very useful and had presented a "public"-centric alternative model and theory. In this chapter, we attempt a reconciliation to explore whether our framework can be accommodated within a particular way of looking at open development.

Open development like open institutions outlines how the network public model can realistically take shape. For open institutions, relevant publics will have full access to an organization's informational resources. Public interest bodies will include relevant publics in their decision-making processes, fulfilling the aspiration of true participatory democracy. Relevant publics will participate in production and distribution of public goods through new networked forms, rendering the processes both more efficient and just. This ideal type of *public network,* with open institutions engaging in structured democratic relationships with *relevant publics*, can anchor itself at different levels: global, national, subnational, and local. It also extends across functional focuses that define different public interest institutions.

Two specific areas are suggested as instantiations for applying the proposed framework of *open development* as open institutions. Community development projects can be evaluated for their *openness* with respect to their communities. New development projects can take into account *open institutional* design principles to develop close, ICT-mediated relationships with their communities. Another area where this concept can be usefully employed is with respect to digital platforms, which are becoming the central infrastructure of digital societies. Such platforms should be evaluated for *openness* on the basis of the criteria we have presented. New platforms should be designed with these criteria in mind, with regulators playing an oversight role.

The proposed open institutional design is an ideal or typical description. It provides a set of standards that can be used to assess the extent of openness of an organization and the degree to which open development is being pursued in a given space. Such judgments would be made against not an abstract notion of complete openness but a contextual one of plausible improvements and evolution.

Open Institutions as Antidote to Institutional Capture

In stressing the networked nature of public interest institutions and their relevant publics, a frequent mistake is to forgo the need for, and benefits of, a certain, continued hierarchical relationship among them. A hierarchy within the network of public interest organizations means that higher layers in the hierarchy provide the outer *constraining rules* for institutions at lower levels. This is like a multiple-shell structure of nested networks, quite like Ostrom's eighth principle for commons management, which is to "build responsibility for governing the common resource in nested tiers from the lowest level up to the entire interconnected system" (Walljasper 2011).

In the network age, the state will retain its central role in the production of public goods and of society-wide coordination. Many functions would, and must, continue to take place at hierarchically higher levels of the *network public*, such as those potentially involving conflicts of interest between relevant publics of different organizations at a lower level or where the benefits of scale are very high. The key democratic principle of subsidiarity will be observed, whereby the rules extended to the next lower layer are to be the minimum required for the latter's effective functioning. Such *outside rules* for public interest organizations will lay the principles for (1) identification of the *relevant public* and (2) how transparency, participation, and collaboration will generally be structured. To leave these tasks entirely to the concerned institution or organization is a recipe for institutional capture. Institutional capture is perhaps the single most significant problem of social organization that we face today. A politically organized society is blamed for a high degree of institutional capture. This justification, although not completely untrue, cannot be allowed to justify an alternative corporatist organization of society, which seems well under way.

Strong institutional improvements are required as we witness the network age assault on the public sector, which is denied its key social coordination role. In pushing back the public sector, or the state, from this role, it is substantially being taken over by transnational corporations. This can be witnessed in the manner in which digital corporations are often seen as providing key public goods, among other things. Big business increasingly seeks self-regulation, which is a nebulous concept, including through employing its global muscle against nationally bound public systems. It has also begun to develop captive *community* and stakeholder groups that provide a veneer of publicness to the neoliberal governance model of multistakeholderism that they promote. The situation produces new kinds of institutional captures that take advantage of the structural and normative fluidities amid shifts from rigid hierarchical to networked institutional and organizational forms. New forms of capture must be analyzed and understood anew. To resist these new age captures, appropriate hierarchically ordered networks rather than complete self-organization of every layer of networked systems becomes fundamentally important.

Public interest institutions have traditionally been designed in an inward-looking manner that is function focused. Public input and participation have largely been limited to elections of public authorities, complemented lately by sporadic processes of participatory democracy. There is a new opportunity in the age of digital networks to employ a fundamentally different design for social institutions. Here, the imperative of participation by and accountability to the *relevant public* is almost as important as the functional purpose of an institution. Better engagement of relevant publics by public institutions can help them effectively determine their agenda and outcomes, not only at macro levels (national, state, and local bodies) but also at meso- and micro- social levels (institutions and organizations).

Although it may appear counterintuitive, such a new participatory focus can also improve institutional efficiency on the premise that the public knows its interest best (the wearer knows best where the shoe pinches). Democratic participation is often seen to connote somewhat chaotic conditions causing harm to efficiency, as expressed in the quip, "a camel is a horse made by a committee" (with great injustice to the natural languid beauty of camels).

An open institutional context, if developed appropriately, may take us beyond such an efficiency versus participation trade-off. This is the core idea behind the peer-to-peer (P2P) movement, best expressed in the great success of free and open source software. Public participation in institutional workings further serves the ideological purpose of ensuring the most equitable distribution of power plausible in the society without compromising its various institutional efficiencies to unacceptable levels. Public institutions will be able to produce and distribute public goods in a much better way, employing the best collaborative possibilities.

Whereas the P2P movement focuses on the economic element of coproduction as the key ingredient of *openness*, we have stressed the political element of codetermination by a community or public of how organizations and institutions function. This is only a matter of different emphasis, since meaningful P2P production also requires codetermination, and codetermination of *open development* would be in vain if it did not lead to real concrete and useful outcomes. The economic and political approaches to openness, respectively, of the P2P movement and our conception of *open development*, are therefore complementary.

A new social arrangement with *open institutions* in intensive interaction with their relevant publics will be considered open not because it minimizes prior community and public rules or institutionalization in favor of flexible *pragmatic* relationships; that is the neoliberal model. On the contrary, it will be open because it is fundamentally designed with an outward orientation for effective control by its relevant publics and against capture by insiders.

Progress toward such an ideal constitutes *open development*. As mentioned, the focus here is not just on institutional accountability; it includes distributed, collaborative mechanisms of production of public and other economic goods, and their equitable distribution. Open markets, from an open institutional perspective, will be an important part of this new ecology. Such markets will be open not because they defy regulation but because they will be framed by collectively developed rules for fair play.

A few caveats are in order. Whether people will actually engage in multiple relevant publics to which they may logically belong will have to be ascertained. The following questions should be kept at the forefront of further research: Would public apathy not be greater rather than less in such complex contexts? What would aid and incentivize participation, and do ICTs have a role to play here? What is the cultural context of such granular participatory democracy? Will the complexity of the new requirements of engagement allow institutions to capture participation and fake legitimacy, while weakening the current state-based controls and accountabilities? Even if lots of people actually engage, what are the trade-offs vis-à-vis the effective operation of an organization in contexts where resources are always limited? Does it at all effectively address the faults of multistakeholderism, where the level of participation can become dependent on the level of resources that one possesses, which tends to skew outcomes?

These are difficult questions on matters of fundamental social importance. For sensitive and important matters, it is wise not to rush to destroy what one is not sure how to rebuild. This is our major concern with regard to the anti-institutional, anarchist tendencies of certain openness proponents. While the weakest and most marginalized people and groups are in the greatest need of transformational change, they also possess the least risk-taking capacity. They most need the support of strong public interest institutions. Efforts to induce change need to be analyzed through robust institutional frameworks, which disproportionately focus on marginalized interests. At the same time, it is important to be bold and counter the neoliberal ideological framing of the pervasive digitally inspired social changes currently under way through alternative theories that are adequate to the novel context. Abdication in this regard can be equally harmful to public interest.

Notes

1. The Open Government Partnership, for instance, lays out basic benchmarks of an open government in its Open Government Declaration (Open Government Partnership 2011).

2. In a 2013 op-ed piece in the *New York Times*, "Open and Closed," Evgeny Morozov quotes Jeff Jarvis, whom Morozov describes as an "Internet pundit," as stating that "owning pipelines, people, products, or even intellectual property is no longer the key to success. Openness is."

3. Since both institutions and organizations can be thought of as a collection of rules, March, Friedburg, and Arellano (2011) discuss the unclear boundaries between the analytical categories of institution and organization, gleaned by examining stable rules and how they change over time. We likewise emphasize the institutional aspects of organizations in terms of their enduring rules over time. Such an emphasis is especially important in these formative times of digital societies, where long-term socioinstitutional designs are currently being set. We place "open development" and "open institutions" in this larger context.

4. See Mungai and Van Belle (chapter 5, this volume) and Moshi and Shao (chapter 12, this volume).

5. For a critique from a privacy perspective, see Malcolm (2016), and for a trade justice viewpoint, see James (2017).

6. We understand that institutions and organizations, especially in complex new contexts, can take up a variety of functions, some more easy to associate with public interest than others. The public interest test in such cases is the element of primacy and whether private interest actions are nested in higher public interest norms or vice versa. It is in this sense that the market, a fair and just one, is a public interest institution, while a huge number of private interest activities are nested within it. But if private interests overwhelm the nature of market relationships, as in institutional capture of the market, one will be unable to keep considering the institution of the market as being in the public interest. Karl Polanyi's concept of a market's embeddedness in social institutions comes to mind here. It is possible for corporate actors to take a series of actions that actually are *primarily* motivated by public interest (the *primacy* element will have to be assessed), and, if these meet other conditions of the definition, they can very well be considered as open development.

7. The state's current political processes may have limitations at the macro level of the government and with regard to public interest articulation. This condition can also be improved by applying the open development concept to the level of government as an institution, toward an *open government*. Such a conception of *open government* would be much wider than the usual application of the term, going beyond transparency to involve thorough participatory democracy and the collaborative production of public goods.

8. The Government of Canada employs the concept of "open by default" in its Canada Action Plan on Open Government (Government of Canada 2014).

9. It is not just those organizations that are solely devoted to the public interest that are considered public interest organizations. See the European Union's definition of a public interest entity at BDO Global, https://www.bdo.global/en-gb/services/audit-assurance/eu-audit-reform/what-is-a-public-interest-entity-(pie). Corporations with publicly traded shares are also considered public interest entities. Any undertaking can be designated a public interest entity depending on "the nature of their business, their size or the number of their employees." In the same way as no property is absolutely private, and public authorities have various kinds of rights to it, every organization has some public interest aspect. Our definition of *open development* will apply to public interest aspects of all organizations.

References

Alibaba Group. 2016. "Fact Sheet: Electronic World Trade Platform." http://www.alizila.com/wp-content/uploads/2016/09/eWTP.pdf?x95431.

BDO Global. n.d. "What Is a Public Interest Entity (PIE)?" https://www.bdo.global/en-gb/services/audit-assurance/eu-audit-reform/what-is-a-public-interest-entity-(pie).

Benkler, Yochai. 2006. *The Wealth of Networks: How Social Production Transforms Markets and Freedom*. New Haven, CT: Yale University Press.

Betancourt, Michael. 2016. *The Critique of Digital Capitalism: An Analysis of the Political Economy of Digital Culture and Technology*. Brooklyn, NY: Punctum Books.

Buskens, Ineke. 2013. "Open Development Is a Freedom Song: Revealing Intent and Freeing Power." In *Open Development: Networked Innovations in International Development*, edited by Matthew L. Smith and Katherine M. A. Reilly, 327–352. Cambridge, MA: MIT Press; Ottawa: IDRC. https://idl-bnc-idrc.dspacedirect.org/bitstream/handle/10625/52348/IDL-52348.pdf?sequence=1&isAllowed=y.

Cañares, Michael P. 2014. "Opening the Local: Full Disclosure Policy and Its Impact on Local Governments in the Philippines." In *Proceedings of the 8th International Conference on Theory and Practice of Electronic Governance*, 89–98. New York: ACM Press.

Caulfield, Mike. 2016. "Putting Student-Produced OER at the Heart of the Institution." *HAPGOOD* (blog), September 7. https://hapgood.us/2016/09/07/putting-student-produced-oer-at-the-heart-of-the-institution/.

Chang, Ha-Joon. 2007. *Institutional Change and Economic Development.* Tokyo: United Nations University Press.

Cyranek, Günther. 2014. "Open Development in Latin America: The Participative Way for Implementing Knowledge Societies." *Instkomm.De.* http://www.instkomm.de/files/cyranek_opendevelopment.pdf.

Davies, Tim. 2012. "What Is Open Development?" *Tim's Blog: Working for Social Change; Exploring the Details; Generally Quite Nuanced,* September 10. http://www.timdavies.org.uk/2012/09/10/what-is-open-development/.

Dewey, John. 1927. *The Public and Its Problems: An Essay in Political Inquiry.* New York: Henry Holt.

Doctoroff, Daniel L. 2016. "Reimagining Cities from the Internet Up." *Medium,* November 30. https://medium.com/sidewalk-talk/reimagining-cities-from-the-internet-up-5923d6be63ba.

Foster, John Bellamy, and Robert W. McChesney. 2011. "The Internet's Unholy Marriage to Capitalism." *Monthly Review* 62 (10): 1–30. https://monthlyreview.org/2011/03/01/the-internets-unholy-marriage-to-capitalism/.

Gaventa, John. 2006. "Triumph, Deficit or Contestation? Deepening the 'Deepening Democracy' Debate." IDS Working Paper 264, July 2006. Brighton: Institute of Development Studies. https://assets.publishing.service.gov.uk/media/57a08c27ed915d3cfd0011e6/gaventawp264.pdf.

Government of Canada. 2014. "Canada's Action Plan on Open Government 2014–16." Treasury Board of Canada Secretariat, Government of Canada. https://open.canada.ca/en/content/canadas-action-plan-open-government-2014-16.

Halal, William E. 2008. *Technology's Promise: Expert Knowledge on the Transformation of Business and Society.* Basingstoke: Palgrave Macmillan.

Harvey, Blane. 2011. "Negotiating Openness across Science, ICTs and Participatory Development: Lessons from the AfricaAdapt Network." *Information Technologies & International Development* 7 (1): 19–31.

James, Deborah. 2017. "Twelve Reasons to Oppose Rules on Digital Commerce in the WTO." *Huffington Post,* December 5. http://www.huffingtonpost.com/entry/twelve-reasons-to-oppose-rules-on-digital-commerce_us_5915db61e4b0bd90f8e6a48a.

Lippmann, Walter. 1927. *The Phantom Public: A Sequel to "Public Opinion."* New York: Macmillan.

Malcolm, Jeremy. 2016. "TISA Proposes New Global Rules on Data Flows and Safe Harbors." *Deeplinks Blog,* Electronic Frontier Foundation, October 24. https://www

.eff.org/deeplinks/2016/10/tisa-proposes-new-global-rules-data-flows-and-safe-harbors.

March, James, Erhard Friedberg, and David Arellano. 2011. "Institutions and Organizations: Differences and Linkages from Organization Theory." *Gestion Y Politica Pública* 20 (2): 235–246.

McChesney, Robert W. 2013. *Digital Disconnect: How Capitalism Is Turning the Internet against Democracy*. New York: New Press.

Michener, Gregory. 2015. "Policy Evaluation via Composite Indexes: Qualitative Lessons from International Transparency Policy Indexes." *World Development* 74:184–196.

Morozov, Evgeny. 2013. "Open and Closed." *New York Times*, March 16. http://www.nytimes.com/2013/03/17/opinion/sunday/morozov-open-and-closed.html?mcubz=0.

Murillo, Felipe. 2017. "When the 'Open Wash' Comes with 'Open Everything.'" *Paris Innovation Review*, June 19. http://parisinnovationreview.com/2017/06/19/open-wash/.

Open Government Partnership. 2011. *Open Government Declaration*. September 2011. https://www.opengovpartnership.org/open-government-declaration.

Rosenberger, Sascha. 2014. "ICTs and Development, What Is Missing?" Institute of Development Research and Development Policy Working Paper 203. Bochum: Ruhr-Universität Bochum. http://www.donorscharter.org/resources/ICT-Development-What-is-Missing.pdf.

Singh, Parminder J. 2017. *Developing Countries in the Emerging Global Digital Order—A Critical Geopolitical Challenge to which the Global South Must Respond*. Bangalore: IT for Change. https://www.itforchange.net/developing-countries-emerging-global-digital-order.

Singh, Parminder J., and Anita Gurumurthy. 2013. "Establishing Public-ness in the Network: New Moorings for Development—a Critique of the Concepts of Openness and Open Development." In *Open Development: Networked Innovations in International Development*, edited by Matthew L. Smith and Katherine M. A. Reilly, 173–196. Cambridge, MA: MIT Press; Ottawa: IDRC. https://idl-bnc-idrc.dspacedirect.org/bitstream/handle/10625/52348/IDL-52348.pdf?sequence=1&isAllowed=y.

Smith, Matthew L., and Laurent Elder. 2010. "Open ICT Ecosystems Transforming the Developing World." *Information Technologies & International Development* 6 (1): 65–71.

Smith, Matthew L., and Katherine M. A. Reilly. 2013a. "Introduction." In *Open Development: Networked Innovations in International Development*, edited by Matthew L. Smith and Katherine M. A. Reilly, 1–14. Cambridge, MA: MIT Press; Ottawa: IRDC. https://idl-bnc-idrc.dspacedirect.org/bitstream/handle/10625/52348/IDL-52348.pdf?sequence=1&isAllowed=y.

Smith, Matthew L., and Katherine M. A. Reilly. 2013b. "The Emergence of Open Development in a Network Society." In *Open Development: Networked Innovations in International Development*, edited by Matthew L. Smith and Katherine M. A. Reilly, 15–50. Cambridge, MA: MIT Press; Ottawa: IRDC. https://idl-bnc-idrc.dspacedirect.org/bitstream/handle/10625/52348/IDL-52348.pdf?sequence=1&isAllowed=y.

Vabø, Mia. 2009. "New Public Management: The Neoliberal Way of Governance." National Research Institute of Social Sciences Working Paper 4. Reykjavík: University of Iceland. http://thjodmalastofnun.hi.is/sites/thjodmalastofnun.hi.is/files/skrar/working_paper_4-2009.pdf.

Walljasper, Jay. 2011. "Elinor Ostrom's 8 Principles for Managing a Commons." *On the Commons*, October 2. http://www.onthecommons.org/magazine/elinor-ostroms-8-principles-managing-commmons.

Reflections II

11 What Makes an Agriculture Initiative Open? Reflections on Sharing Agriculture Information, Writing Rights, and Divergent Outcomes

Piyumi Gamage, Chiranthi Rajapakse, and Helani Galpaya

Introduction

Agricultural initiatives in the development sphere have seen torrid evolution. The Green Revolution conjures up images of scientists in lab coats coming up with efficient ways to grow staple crops such as wheat and maize, which were then transferred to developing countries in a wave of technocratic initiatives between the 1930s and 1960s. Lewis' (1954) economic development model suggested that the ultimate goal for the process of economic expansion should be to see complete absorption of small and subsistence rural farming by the capitalist sector. Yet, through implementing the Green Revolution, development practitioners and policymakers soon realized that including Indigenous and rural subsistence farmers in policy and planning could actually be an important development objective in and of itself (Parnwell 2008).

Over the course of the next four decades, putting the perspectives of rural and poor farmers first became important for generating sustainable livelihoods that were capable of dealing with increasing pressures on the environment and higher rates of rural to urban migration (Chambers 1994). Nevertheless, there remains a tension between agriculture initiatives imposed from the top and those that stress pluralist approaches to empower farmers. A review of public sector agricultural extension initiatives in developing countries by Rivera, Qamar, and Van Crowder (2001), for instance, demonstrated mixed results, recommending increasing partnerships between farmers and supportive agricultural organizations and businesses, greater decentralization toward lower levels of government, and subsidiarity at the grassroots level.

Open processes might facilitate a more integrated model for agricultural initiatives, but there has not been much research into this. Smith and Seward's (2017) practice-based view of openness potentially enables multidirectional information flows between producers, distributors, and consumers of information. Sharing agricultural information openly implies that farmers can use or repurpose that information to meet their own needs. Likewise, enabling farmers to contribute their own knowledge via a crowdsourcing platform could unlock a channel that centralized agricultural knowledge banks can use to improve, test, and enrich their databases. However, as Dearden, Walton, and Densmore (chapter 8, this volume) argue, just because an organizer builds open processes into an initiative does not mean that farmers will learn how to participate and benefit from the initiative in the same way. This discussion reflects on the manner in which open agricultural initiatives are conceptualized and enacted. Our contribution is to encourage a notion of open agriculture that is rooted in developing the *writing rights* of farmers.

Dearden, Walton, and Densmore's (chapter 8, this volume) theoretical framework offers a unique and critical lens through which both the design of agricultural initiatives and the outcomes experienced by farmers may be investigated. The concept of *writing rights* centers the investigation on literacy practices that farmers use not only to receive and adopt standard agriculture information but also to contribute their own knowledge and experience within the initiative through *writing relationships*. Examining the literacy practices and situated activities of farmers simultaneously provides an avenue for understanding agricultural practice and open process participation. It also enables fuller and richer descriptions of the everyday practices and challenges that farmers confront while working toward their agricultural goals. Farming practices related to pest control, crop rotation, planting, and harvesting can vary substantially across regions, even within the same country. This knowledge is situated in the informal learning practices of farmers, which allow us to begin to understand the divergent outcomes of agricultural initiatives.

Coming to Grips with Writing Rights in Open Agricultural Initiatives

Our reflections stem from research that began after the Sri Lanka Standards Institution defined good agricultural practice (GAP) standards for specific crops targeted for export to European markets (de Zoysa 2016). When the GAP standards were released, farmers could only access the information by talking with agricultural extension officers. They could also call 1920, a crop advisory service run by the Department of Agriculture, or look at print copies of the GAP standards while visiting the department. In practice, very few farmers actually looked at print copies of the information, as the majority of farmers had no interest in engaging with the highly technical information. They wanted access to the practical information that was lacking in the standards. Thus, farmers who did not learn how to follow the new standards were excluded from exporting their products to foreign markets, thereby facing significant economic consequences.

In 2015, LIRNE*asia* began working with the Department of Agriculture to create the standards for three crops in the cucurbit family—snake gourd, bitter gourd, and luffa—that would be available to farmers via a smartphone application. LIRNE*asia* hoped that making the information more openly accessible would facilitate uptake of the GAP standards, improving export rates and the income levels of farmers. The project, titled *Inclusive Information Societies 2—How Knowledge, Information and Technology Can Connect Agricultural and Service-Sector Small Producers to Global Supply Chains* (LIRNE*asia* 2014), aimed to assess the impact of sharing this information freely through a mobile app designed for farmers. However, the initial design of the app was focused on sharing information, as the project was largely concerned with the impact of this information on increasing exports.

In contrast, when we contemplated how Dearden, Walton, and Densmore's (chapter 8, this volume) framework might be applied to our initiative, we needed to confront some of our ideas and assumptions of what an open agriculture initiative entails. We initially considered sharing information through the mobile app as open data, especially since we planned to share the app with the public for free once the main project finished. However, *writing rights* implies that open processes enable farmers to have a voice, to reuse and repurpose the information, or to contribute through their own experience and knowledge. Thus, we decided to explore how more active open practices (rather than merely consulting information) could be integrated into the project.

The difficulty with changing the design of the app to provide opportunities for writing relationships to occur was that there were many different actors, such as farmers, extension workers, project coordinators, and the

Sri Lankan Department of Agriculture, who were all engaged in the initiative for different reasons. For example, the Department of Agriculture was focused on increasing cucurbit exports, while farmers were participating in the initiative in divergent ways. Drawing on the research of Karanasios and Allen (2013) and Karanasios and Slavova (2014), Dearden, Walton, and Densmore (chapter 8, this volume) outline how activity systems of diverse actors run in parallel. Sometimes activity systems interact symbiotically, generating significant value, whereas at other times they may cause tension and conflict. It is important to identify activity systems independently, as well as how they interact with each other. Considering the farmers' wider information-seeking activities in relation to the Department of Agriculture's services and objectives was key to understanding how and why activity systems differed between these actors.

We learned a great deal from observing the situated activities of service agents working for the Department of Agriculture's centralized advisory service, called 1920, which is a call-in service available to farmers seeking information. Four or five agents sit in a common room to answer the farmers' inquiries. Whenever a call comes in, the farmer's basic information is recorded in a central database. To answer the question, the agent consults a set of books lying on a common table or refers to specific guidelines provided by the Department of Agriculture. Agents also frequently use their tacit knowledge to answer questions directly. When the agent cannot answer the query immediately, the caller is directed to a research institute with greater specialized knowledge. However, the central database did not facilitate allowing the service agents to track whether questions had been answered and by whom, so this information was not gathered.

When we spoke with seven farmers during two focus groups, we confirmed that they valued direct responses to their problems. They indicated a low level of trust in the 1920 service. Most farmers were accustomed to solving problems through personal networks, family wisdom, and in-person consultation with extension workers. Farmers avoided the 1920 service because of frequent busy signals or being redirected to two or three different places without receiving satisfactory responses to their questions.

The farmers also demonstrated a strong willingness to contribute actively to improving the materials and services provided by the Department of Agriculture. For example, farmers questioned why outdated and impractical fertilizer recommendations were still in the GAP guideline documents. Participants argued that fertilizer requirements change according to regional land conditions, and the suggested fertilizer amounts did not work in their area. They insisted that their contextualized knowledge needed to be taken into consideration when designing materials to convey specific advice or recommendations. Other participants stated that they had successfully modified and used knowledge from other countries on their own farmland. These conversations made us realize the importance of the farmers' writing rights.

This deeper understanding of the social contexts and interacting activity systems enabled us to respond to the needs of the farmers and the Department of Agriculture by developing features for the app that harnessed open practices in ways that were consistent with their customary interactions and that improved the effectiveness of the service. We designed new features for the app that reflected the rules and procedures set in place by the 1920 service. The new features enabled users to send a question to a designated official at the Department of Agriculture, who would then respond to the queries. The service agents likewise stated that in many cases it was impossible to give sufficient answers to farmers' questions because the farmers could not adequately convey all the necessary information to identify problems by simply describing them over the telephone.

We added a feature to take a photo of the problem to accompany questions sent to the service. Figure 11.1 shows how farmers can attach a photo or voice recording to their question. The agent who received the questions would either send an answer via a typed message or would call back and speak to the farmer. We also added conversation threads because there was no method for the farmers to know which 1920 agent had phoned them. We introduced message threads so that previous messages could be tracked and farmers could carry on the conversation with the department until they got a satisfactory answer.

Ultimately, our position as a mediator among the actors helped to solidify the success of the initiative and the implementation of writing rights in culturally relevant ways, and to support a sustainable future for the initiative. Fashioning the app according to the activities taking place within the 1920 service enabled a quick handover, with little need for learning by the service agents. This enabled a context-driven approach to open process integration, which was designed to facilitate more active contributions by the farmers.

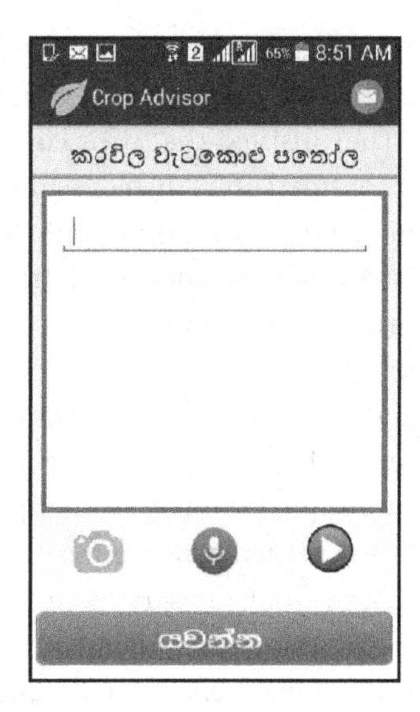

Figure 11.1
Screenshot of the window farmers use to send questions to 1920.
Source: Mobile app designed by LIRNE*asia*.

Moving Beyond Divergent Outcomes of Open Agriculture Initiatives

While we were successful in adding more opportunities for farmers to contribute their knowledge and experience to the standardized initiative, the farmers experienced divergent outcomes unrelated to the initiative's objectives. We interviewed five farmers who had used the app for three months, and all demonstrated a range of literacy practices, moderated partially by personal characteristics and circumstances. Some younger farmers were dependent on extension officers and did not feel confident about finding appropriate information through the app, whereas older farmers were interested in exploring the app but did not need to use it, probably because of their sound knowledge of agricultural practices.

Moreover, some farmers spoke about how important it was that the app helped them build new contacts, perhaps a reflection of the high value that is placed on social interactions in the farmers' communities. All the farmers using the app began conversations with the Department of Agriculture. This has increased farmers' trust in 1920. It has also created a conversation among farmers about sharing knowledge and providing feedback to the Department of Agriculture. Also notable was the way in which certain farmers who learned to use the app were quickly able to offer insights on how to improve its features.

Of equal interest to our project team, and to the Department of Agriculture, was how using the app and participating in a stronger network surrounding cucurbit farming might improve yields and exports to foreign markets. Unfortunately, it was too early to tell how and whether using the app was affecting farming practices. Most of the farmers were not growing the focal crops at the time of study. However, since farmers were increasing communication with the Department of Agriculture by sharing photos and seeking help for the crops they were growing, the improved 1920 service may help the Department of Agriculture learn more about local farming patterns and concerns.

Furthermore, Dearden, Walton, and Densmore's (chapter 8, this volume) lens emphasizes a broader meaning of open development that surpasses immediate agriculture objectives and encompasses the self-determination, resources, and active participation of individuals in open initiatives. It is not sufficient to consider only the implementation of a smartphone app, which means engaging with the ways in which this medium is used to claim writing rights.

The cucurbit farmers' situation reflects that of many Sri Lankan farmers, who are claiming their agency and challenging the farmers' position within national agriculture initiatives such as the GAP standards framework. The confidence and self-sufficiency they demonstrate is a result of their longstanding traditions, strong social and professional networks, and resources they have to respond to local environmental and market conditions. Challenging the information provided within the GAP standards, and the propensity farmers had to solve their own problems using a range of resources at their disposal, has likely resulted in some improvements in their relationships with the Department of Agriculture and in their farming results.

However, the Department of Agriculture and its cooperating organizations are the ones influencing how discourses of openness and agriculture development are woven together within their collaborations with the farmers. Open development may provide a better conceptual framework and program design for these institutions to follow. Yet, farmers' writing rights within such a framework may only reflect the boundaries set in place by the initiative. Perhaps the most important takeaway we gained from Dearden, Walton, and Densmore's (chapter 8, this volume) theoretical lens is that using open processes as a means of treating the farmers as active contributors, enabling their use of material and social resources, and furthering their engagement in the initiative is a first step toward positive social transformation. Tackling the broader issues of institutional reform and the value of farmers' writing rights within the Department of Agriculture's institutional structures is the next step toward attaining meaningful transformation through open development.

References

Chambers, Robert. 1994. "Foreword." In *Beyond Farmer First: Rural People's Knowledge, Agricultural Research and Extension Practice*, edited by Ian Scoones and John Thompson, xiii–xvi. London: Intermediate Technology Publications.

de Zoysa, Nirmala. 2016. "GAP Adoption and GLOBAL GAP Certification." Summary Report. https://gapcambodia.files.wordpress.com/2016/12/sri-lanka-nirmala-de-zoysa-summary2.pdf.

Karanasios, Stan, and David Allen. 2013. "ICT for Development in the Context of the Closure of Chernobyl Nuclear Power Plant: An Activity Theory Perspective." *Information Systems Journal* 23 (4): 287–306.

Karanasios, Stan, and Mira Slavova. 2014. "Legitimacy of Agriculture Extension Services: Understanding Decoupled Activities in Rural Ghana." Presented at *European Group on Organization Studies, 30th EGOS Colloquium*, Rotterdam, The Netherlands, July 3–5, 2014. https://www.researchgate.net/publication/272681221_Legitimacy_of_agriculture_extension_services_Understanding_decoupled_activities_in_rural_Ghana.

Lewis, W. Arthur. 1954. "Economic Development with Unlimited Supplies of Labour." *The Manchester School* 22 (2): 139–191.

LIRNE*asia*. 2014. *Inclusive Information Societies 2 (IIS2)—How Knowledge, Information and Technology Can Connect Agricultural and Service-Sector Small Producers to Global Supply Chains*. Colombo: LIRNE*asia*. http://lirneasia.net/wp-content/uploads/2015/04/LIRNEasia-Inclusive-info-societies-2015-17-proposal-v8_FINAL.pdf.

Parnwell, Michael J. G. 2008. "Agropolitan and Bottom-Up Development." In *The Companion to Development Studies*, 2nd ed., edited by Vandana Desai and Robert B. Potter, 111–114. London: Hodder Education.

Rivera, William M., M. Kalim Qamar, and Loy Van Crowder. 2001. *Agricultural and Rural Extension Worldwide: Options for Institutional Reform in the Developing Countries*. Rome: Food and Agriculture Organization of the United Nations. http://www.fao.org/3/a-y2709e.pdf.

Smith, Matthew L., and Ruhiya Seward. 2017. "Openness as Social Praxis." *First Monday* 22 (4). https://firstmonday.org/ojs/index.php/fm/article/view/7073.

12 Using the Critical Capabilities Approach to Evaluate the Tanzanian Open Government Data Initiative

Goodiel C. Moshi and Deo Shao

Introduction

The framework proposed by Zheng and Stahl (chapter 9, this volume) challenges the idea that open development initiatives have positive and uniform objectives. They suggest that open data initiatives are influenced by *ideological* and *political* structures and that progress and outcomes should be determined based on how the initiative affects the *well-being* and *agency freedom* of those it intends to support. It assumes that actors will have different capabilities and interests in engaging in open initiatives and that power differentials affect how and why actors get involved.

For open data initiatives, it is still not clear how the critical capabilities approach (CCA) applies. Its four pillars—human-centered development, protection of human agency, human diversity, and democratic disclosure—can be interpreted both within and outside the institutions, driving open data initiatives quite differently. This is why the critical questions provided by the evaluation framework make it easier for researchers and practitioners to target how and why the principles translate into the structures, processes, and outcomes of the initiative over a project's life cycle. This reflection considers our experience researching Tanzania's open government initiative (TOGI), focusing specifically on the open data initiative, to interrogate the effectiveness of Zheng and Stahl's CCA evaluation framework.

TOGI was a well-established open government initiative. It began in 2011 when Tanzania joined the Open Government Partnership (OGP). The overarching goals of the OGP are to promote transparency, accountability, and citizen participation. As part of TOGI, Tanzania carried out two action plans. Action Plan I occurred in 2012 and 2013, with the goal of establishing the initiative across public sector institutions (United Republic of Tanzania 2012). In 2014, the government crafted Action Plan II, which lasted until 2016 (United Republic of Tanzania 2014b).

During the implementation of Action Plan II, the government deployed a central open data platform (www.opendata.go.tz), where it began to publish open government data. At the time of writing, the portal had over 186 data sets from the health, education, national statistics, and water sectors, published in CSV format, but it is now closed. The education sector, in particular, produced almost half the open data (eighty-five data sets overall). For this reason, when we began researching TOGI's open data initiative, we expected to see a wide range of progress and experience in the education sector. After all, education data sets are the most popular and frequently downloaded data sets on the Basic Statistics Portal. Theoretically, it should have been easy for us to identify a number of actors benefiting from this initiative. However, evaluating the initiative using the CCA framework depicted a different picture of progress. Numerous problems were uncovered, particularly when examining the design and implementation of TOGI's open data initiative within the education sector.

After our research ended, Tanzania announced its withdrawal from the OGP in November 2017 (OGP Support Unit 2017). Ultimately, our evaluation of the open data initiative in the education sector using the CCA points out some of the reasons for this outcome. Had the main decision makers used the CCA framework to guide the design and implementation of the initiative, perhaps we would be witnessing a different result. In this chapter, we explore our research conducted in conjunction with the Tanzanian government, the World Bank, and civil society representatives to outline the main design features and implementation challenges of the initiative according to the CCA. We then compare our evaluation with that of the OGP upon completion of Action Plan II.

TOGI's Open Data Design within the Education Sector

A host of actors were involved in the design of TOGI's open data initiative, including the National Bureau of Statistics (NBS), the Ministry of Education (MoE), the World Bank, and Tanzania's e-Government Agency (eGA). The eGA is tasked with coordinating and providing oversight as well as enforcing e-government standards in the public service. A consultative committee led by the NBS and facilitated by the World Bank was primarily responsible for making decisions regarding the open data initiative. As one senior MoE staff member stated, "I was one of the founding members of the TOGI committee. Our committee strategized the initial setup of the initiative, including prioritizing dataset categories for open publication." The committee created both action plans, which laid out the design of the initiative across all government institutions, including the MoE. The roles and responsibilities of the stakeholders also revolved around the outputs and procedures outlined in Action Plan II (United Republic of Tanzania 2014). The context and structure of the initiative, as well as the concept of openness enacted, follows almost entirely from the action plans.

Action Plan II concentrated on releasing preexisting government information to the public in open data formats. The primary educational open data producers are the MoE and the NBS. The action plan also listed the National Examination Council of Tanzania, a separate government institution that has released information on a different public website but not in conjunction with the Basic Statistics Portal. All these government institutions have information that had not been shared publicly. For instance, the NBS had collected data through household surveys to compile representative statistics about Tanzania's demographics since 2012. Household questionnaires collected information about the level of education people have, which can be compared with other educational data for planning and management. The MoE, likewise, has national records containing details of Tanzania's educational system, such as the number of schools in a district, male and female enrollments, student-to-teacher ratios, and performance metrics. The rationale underpinning openness in this instance is that government institutions have this information and it is in the public's interest that it be shared (United Republic of Tanzania 2014).

Furthermore, two of the four objectives of Action Plan II for the open data initiative focused on institutionalization of the initiative to (1) establish a coordinating body and (2) develop supporting guidelines for participating institutions to follow (United Republic of Tanzania 2014, 5). However, there was no specific information regarding the institutional bodies implicated, which created considerable ambiguity. Upon speaking with representatives of the NBS and the MoE, we learned that they had not discussed what an institutional framework might entail, nor how open data responds to their objectives for the education sector. Thus, we were unable to answer the question posed by the CCA regarding the intended consequences of the project (Zheng and Stahl, chapter 9, this volume).

However, because the main focus of the initiative was on releasing preexisting data, data producers participated in workshops regarding data curation. They did not address other matters, and they considered their actions adequate for the time being. None of the representatives we spoke with could give specific examples of how to connect the open data initiative to their MoE objectives. For example, the MoE's mission is to strengthen educational institutions and procedures that will enable Tanzanians to receive a quality education. Instead, the representatives gave general examples of how open data can monitor quality and stimulate discussion in the public sphere. For example, one MoE officer believed that "open data will help the public answer many of their questions regarding trends in education performance." Clearly, the individual's attitude toward the initiative was that open data is a standalone resource that has no immediate bearing on the education sector's policies or practices.

Prior to releasing the third action plan (United Republic of Tanzania 2016a), a group of civil society actors were included in invitation-only consultation meetings. One such group, Twaweza, has taken a key role in open educational data for more than a decade and was invited to these meetings by the OGP. Twaweza lowers the barriers to using open data by developing infographics and mashups to aid in their public advocacy work. It also provided suggestions to make content more easily understandable to citizens by releasing summaries of data in the local language and by providing offline access. However, since Tanzania withdrew from the OGP, it is not clear whether they will ever release or pursue the third action plan.

Overall, the open data initiative began in Tanzania, as it did in many other countries, through participation in the OGP. The definition of openness was imposed on the Tanzanian government, and many of the representatives spoke of the benefits of the initiative mainly in an abstract manner. For instance, all the overarching objectives of the initiative mentioned by government representatives revolved around OGP objectives, such as increasing government transparency, accountability, and citizen participation. None of the representatives could translate these objectives into intended consequences within the education sector specifically. This indicates a lack of ownership of the initiative and may indicate that the decision makers did not have the skills to adapt the program to the Tanzanian educational context.

Implementation Challenges

Many of Zheng and Stahl's (chapter 9, this volume) critical questions for the implementation stage were difficult to answer because we found that the basic preconditions for the initiative were not being met. The main decision makers did not follow through on the day-to-day activities, so there were very few representatives within the MoE or at the NBS who could reflect on the initiative's progress related to more substantive concerns, such as what individuals or groups were the intended beneficiaries or what the intended impacts of the initiative were.

Both the MoE and the NBS had not instituted human resource responsibilities and procedures for the initiative. Although government employees had been assigned to implement the open data initiative, open data responsibilities had not been incorporated into the officers' job descriptions. As one government employee asserted, "There is no one with the sole responsibility for the initiative. It is done on an ad hoc basis. Even though I have taken part in the initiative, my open data activities are not part of my job description and do not appear in my OPRAS [Open Performance Review and Appraisal System] evaluations." Responsibilities had not been included within the strategic plans of government units responsible for producing open data either. This meant that until February 2018, data curation and publication had been done in batches on three occasions. Each time, the World Bank, as an external facilitator, had to request the updates in order for them to happen.

In order to understand who the beneficiaries of the initiative were meant to be, we consulted the action plans. Action Plan II outlined three target groups for TOGI: academics, civil society, and media practitioners. However, these actors may not be the most relevant beneficiaries for the education sector, as other target groups, such as school administrators, other parts of government, regional commissioners, parents, or students, might find more benefit from the initiative. Nevertheless, we did interview academics and journalists engaging with the education sector. The interviews were done between November 2016 and March 2017, and each interview took about ninety minutes. These stakeholders were either unaware of the open data initiative or had not yet used the data it released in their research or media publications. We were at least able to understand the research problems and educational topics these stakeholders were addressing, and thus what kinds of data might be useful to them.

When we asked NBS and MoE representatives whether they had contemplated any user-focused work, they simply acknowledged that for the data to be useful to the intended users they should have some background in statistical skills. There were no plans to help users gain technical and statistical skills to engage with the open data. Instead, the eGA concentrated on a technical solution to support end users by modifying the platform to present data in both simplified and refined ways to allow novice users to engage more easily with the content. They worked closely with other stakeholders, such as the World Bank, to define custom visualizations that they presumed suited user demands. However, we did not collect any significant evidence that the eGA actually understood what the users' needs were in the education sector as the visualizations devised did not respond to any objective that the academics and media practitioners desired for their work.

We also considered that intermediary actors might be better positioned to help end users benefit from the initiative. We spoke with two organizations focused on open data engagement. Tanzania Data Lab (https://dlab.or.tz/) provides training to build capacity in data publication and usage. However, they do not engage stakeholders in the education sector because the lab's manager stated, "These are areas where the funder wants us to focus: water, agriculture, and gender." In contrast, Data Zetu (now managed by the Tanzania Data Lab; https://medium.com/data-zetu) aims to collect data directly from the community and creates tools to use that data to serve that community. The Data Zetu director stressed that, "We are an independent entity volunteering to build tools for [the] community from their data. We are not constrained by any national open data policy or strategy." Data Zetu's mandate is to respond to aspects that have been missing from within the government's initiative, and it wants to remain independent. Nevertheless, the work of both institutions could help beneficiaries gain open data skills. They simply do not target stakeholders within the education sector.

Perhaps the most important benefits of the initiative for the education sector were experienced by government officials and politicians external to the MoE and the NBS. As a World Bank representative pointed out, "Government is not one, but constituted of a number of integrated ministries, departments, and agencies where, at a point in time, an officer in one unit may have a hard time getting data from another unit. Thus, there are situations where government [representatives] become users of open data." For instance, NBS staff mentioned that open data was being used for setting budgets, and a member of parliament had used data to make a case for his or her constituents. Here, open data is clearly benefiting government objectives, such as increased efficiency and responsiveness, which may ultimately indicate a more realistic design for the initiative.

The last missing precondition that presented a significant barrier to the implementation of the initiative was the fact that the MoE did not have a single legal instrument to guide the initiative's institutionalization process. Each government institution involved in TOGI's consultative committee had detailed mandates prior to the advent of the open data initiative. GeoPoll, for example, was subject to laws that prohibited publishing data that were not nationally representative (The Citizen 2017). Different institutions also had different interpretations of applicable laws, in particular the Statistics Act, 2015 (United Republic of Tanzania 2015) and the Access to Information Act, 2016 (United Republic of Tanzania 2016b). Developing a legal framework to facilitate the implementation of TOGI therefore required that the MoE participate in and negotiate at an interdepartmental discussion, where concessions were likely to be made because of the institutional complexity involved.

Another round of policy discussions happened after collecting public opinion regarding the *Open Data Policy Draft* in April 2016. The official policy has not been published at the time of writing and is expected to provide clearer direction regarding the conflicting laws in place. However, once again, mainly the data producers were involved in this discussion. It is also not clear whether the government will pursue the open data initiative, since Tanzania announced its withdrawal from the OGP in November 2017.

Comparing TOGI's End-of-Term Evaluation with the CCA Evaluation

The OGP implemented an independent reporting mechanism (IRM), which evaluated TOGI's progress on its commitments for the period covering Action Plan II (Tepani 2017). The IRM evaluation analyzed the language of the commitments found within the action plan, linked each commitment to an overarching OGP objective, analyzed the potential impact of the commitment, determined whether the commitment was completed, and addressed a final criterion called "Did it open government?" An exemplary commitment was defined as being specific, related to one of the overarching OGP objectives, having the potential to be *transformative* (unclear what this actually means), and near completion. Regarding the question of whether the commitment opened government, the criterion was whether government practice had been stretched beyond business as usual. Information was not provided concerning how *changes* to business as usual were investigated.

According to the IRM evaluation, TOGI's open data initiative was one of two (out of five) initiatives that received ratings of substantial completion and having major impacts on opening government. The rest of the TOGI initiatives received ratings of limited completion and marginal impact on opening government. The open data initiative achieved substantial achievement primarily because of the number of data sets that were published on the Basic Statistics Portal and because a coordinating body had been announced but not yet implemented. Moreover, the initiative scored highly for its impact on open government because this aspect was evaluated according to increasing access to information (instead of the two other pillars, those of accountability or citizen participation). It seems business as usual had changed because one civil society organization representative claimed that the initiative had improved the way that citizens viewed the work of the government.

Most of the recommendations made by the IRM evaluation were concerned with increasing engagement with citizens. However, the initiative was not evaluated on the basis of citizen participation. There was also no mention of the practical steps involved in making the initiative more responsive to citizens by integrating the initiative into institutions like the MoE and the NBS. The evaluation points to Internet penetration rates and the skills citizens need to engage with open data as the main barriers to confront. It does not address any of the internal barriers to implementation that we encountered.

In contrast, applying the CCA evaluation framework involved a more thorough investigation and different sets of progress indicators. The CCA emphasized the lack of a supportive political culture and the failure of the decision makers to establish responsibilities and procedures within the relevant institutions. The way that the CCA links the project (design, implementation, and evaluation) and its principles (human-centered development, human diversity, democratic discourse, and protecting human agency) encourages evaluators to pose holistic, critical questions. In contrast to the IRM evaluation, the CCA does not focus on a predefined set of commitments, instead putting the principles first. This means that we could not ignore the ideological or political constraints, or the lack of beneficiary identification or involvement, because these were not the main commitments at the time. It is simply not possible to conclude that freedoms were enhanced substantively through the publishing of open data sets. Thus, the CCA requires and encourages a richer, deeper analysis of impact.

There is no doubt that the CCA framework was helpful in analyzing Tanzania's open data initiative; however, implementing the evaluation was complicated. Part of the challenge is that the initiative is comprised of a large number of stakeholders, including ministries and agencies, the NBS, the eGA, the World Bank, and civil society organizations. Such a large number of stakeholders, which is common in most open data initiatives, poses a challenge for applying the CCA framework because of the initiative's multisectoral agenda with many competing interests.

Focusing on the education sector enabled us to reduce the complexity of the initiative so that the CCA was more practical. It forced us to take our investigation beyond the immediate actors and to consider a plethora of potential users of open data in education. However, the framework also turned up more important findings that signaled conceptual problems within the initiative. Critical questions related to the definitions of openness enacted, where the openness arises, and who the main decision makers are turned up some of the deeper issues related to how and why the Tanzanian government may have been losing interest and eventually decided to withdraw from the OGP.

References

The Citizen. 2017. "NBS Warns Geopoll against Releasing of Unofficial Statistics." *The Citizen*, August 7. http://www.thecitizen.co.tz/News/NBS-warns-Geopoll-against-releasing-of-unofficial-statistics/1840340-4048486-h1n3drz/index.html.

OGP Support Unit. 2017. "Tanzania Withdraws from the OGP." Open Government Partnership, September 28. https://www.opengovpartnership.org/news/tanzania-withdraws-from-the-ogp/.

Tepani, Ngunga Greyson. 2017. "Tanzania: 2014–2016 End-of-Term Report." Open Government Partnership. https://www.opengovpartnership.org/wp-content/uploads/2001/01/Tanzania_EOTR_2014-2016_for-public-comment_ENG.pdf.

United Republic of Tanzania. 2012. "Tanzania Open Government Partnership (OGP) Action Plan 2012–2013." Open Government Partnership. https://www

.opengovpartnership.org/wp-content/uploads/2019/06/OGP-ACTION-PLAN-REVISED-ON-26-3-2012-1.pdf9__0.pdf.

United Republic of Tanzania. 2014. "Tanzania Open Government Partnership (OGP) Second National Action Plan 2014/15–2015/16." Open Government Partnership. https://www.opengovpartnership.org/wp-content/uploads/2017/05/Tanzania_EOTR_2014-2016_for-public-comment_ENG_0.pdf.

United Republic of Tanzania. 2015. Tanzania Statistics Act, 2015. https://tanzlii.org/tz/legislation/act/2015/9-0.

United Republic of Tanzania. 2016a. "Tanzania Open Government Partnership (OGP) Third National Action Plan." Twaweza. https://www.twaweza.org/uploads/files/FINAL%20OGP%20ACTION%20PLAN%20III%2030_8_2016.pdf.

United Republic of Tanzania. 2016b. The Access to Information Act, 2016. Freedominfo.org. http://www.freedominfo.org/wp-content/uploads/Tanzania-Access-to-Information-Act-2016.pdf.

13 Three Problems Facing Civil Society Organizations in the Development Sector in Adopting Open Institutional Design

Caitlin M. Bentley

Introduction

Over the past forty years, one enduring source of criticism of the contributions of civil society organizations (CSOs) in the development sector has been their increasing neoliberalization as alternative and more efficient channels of development aid delivery (Desai and Imrie 1998; Hulme and Edwards 2013). Wallace and Porter (2013) went so far as to argue that this trend has created a crisis of representation. It is a crisis because CSOs were often considered to be working in the public's interest because of their geographic proximity to their relevant publics, their use of alternative participatory methods, and their public advocacy work (Bebbington, Hickey, and Mitlin 2008; Lewis and Kanji 2009). Yet, the neoliberal approach to development has increasingly dominated CSO practice over the decades, encouraging *managerial* forms of program design, implementation, and evaluation (Eyben 2013). This mode of development practice is not seen as being compatible with representing a relevant public's interest, thus inducing a crisis of representation.

The more powerful and standardized managerial approaches become, the easier it is for researchers and practitioners to accept that there can be no alternatives. Collectively, we have been driven to think about development projects as a series of activities listed in a document table, along with bubble diagrams representing some abstract *theory of change*, without questioning it further.

This is not the reality that Singh, Gurumurthy, and Chami (chapter 10, this volume) subscribe to by any means. They present a new *open institutional design* that could effectively resolve the crisis of representation that development CSOs face. Open institutional design encourages "use of ICTs for institutional redesigning to bring about structural changes that enhance transparency or information sharing, participation, and/or collaboration, in a manner that is primarily motivated by, and contributes to, public interest" (Singh, Gurumurthy, and Chami, chapter 10, this volume). Other scholars have argued for such a refocusing of CSO governance structures to reflect Singh, Gurumurthy, and Chami's main concerns (Banks, Hulme, and Edwards 2015). The difference is that open institutional design offers an innovative method for CSOs to achieve it. Singh, Gurumurthy, and Chami (chapter 10, this volume) suggest that new forms of information and communication networked practices and infrastructure, such as crowdsourcing, peer production, or distributed decision-making, as well as a "multiple-shell structure of nested networks," will provide the necessary skeletal structure to implement effective participatory democratic governance on a potentially massive scale. The proposed open institutional design could revolutionize CSO practice and governance in the development sector.

However, my reflection is about the current reality development CSOs face and what must be confronted as they work toward adopting open institutional designs. I primarily draw on two ethnographic case studies of relationships between bilateral donors and development CSOs in southern Africa, Togo, Canada, and the United Kingdom (Bentley 2017). Within the preliminary stages of this research, I also conducted interviews with eight bilateral and multilateral donors and seven international CSOs based in the United Kingdom and Canada. This research focused on the biggest challenges these organizations faced in carrying out their work, in collaborating, and in using technology for knowledge sharing and accountability. This reflection outlines three problems for consideration to further develop the practical applicability of Singh, Gurumurthy, and Chami's open institutional design.

Problem 1: Identifying a Public Is One Thing, but Engaging a Public through ICT Can Be a Massive Headache

The notion that CSOs should focus their ICT-engagement (information and communications technology) strategies on relevant publics is powerful for more reasons than those discussed by Singh, Gurumurthy, and Chami (chapter 10, this volume). In my research, I found that CSOs spent a great deal of time, energy, and money on disseminating information via websites, e-newsletters, and social media because managers believed that making the information publicly available strengthened accountability toward organizational missions. However, many CSO staff members found such tasks tedious, boring, and of little value. As one program manager stated, "All those sleepless nights, and for what? 100 hits!" Her organization likewise used web and e-newsletter analytics software and commentary web forms to understand who their audience was, but these methods were not adequately informative or reliable. When practitioners did not know who they were engaging with, they tended to value ICT-based activities less. Singh, Gurumurthy, and Chami's suggestion to focus first on identifying relevant publics ensures that practitioners focus more on developing relationships with beneficiaries and less on disseminating information to a universal audience.

That said, Singh, Gurumurthy, and Chami's (chapter 10, this volume) suggestions to identify relevant publics through self-identification or rules-based exclusions are good first steps, but moving beyond this will require greater reflection for two reasons: the complexities involved in reaching out to relevant publics through ICTs and the administrative capacities needed to understand the management of private data. First, a relevant public may not be a homogeneous group, and there will still be a need to develop differentiated strategies to employ ICTs to engage a relevant public in decision-making, participation, and/or collaboration to address the various modes of communication used by different members of the relevant public. Some strategies may require more time and effort to include all sections of a relevant public equitably, and being clearer about who belongs to a relevant public does not resolve difficulties in engaging less powerful members of a public.

For example, in my research with Gender Links (http://genderlinks.org.za/), a regional CSO in southern Africa that is dedicated to increasing gender equality, one of their programs focused on gender mainstreaming in local government. In this case, the relevant public might be citizens in a particular region. Yet, there are also collaborating actors who are integral to the program, such as national and local government representatives, partner organizations, and consultants. Let us assume that all these actors share the same public interest in making local government and its services gender responsive, and thus they may all be considered part of the relevant public. Now consider that each of these sections of the relevant public has different positions and levels of power as well as different ICT access and communication cultures. Differences occur not only between different actor groups but within them as well.

In Mozambique, the Gender Links program officer needed to communicate with local governments via fax, telephone, or email, depending on what these actors had available. In Madagascar, an intern called over two hundred citizens to ensure that information was reaching them. He also had to meet with the national government representative regularly in person. The intern could not call or email the representative because, in the representative's culture, the intern would not be taken seriously. These heterogeneous access and communication patterns make it difficult for CSOs to employ ICTs to engage a relevant public. It is necessary not only to identify who a relevant public consists of but also to develop an understanding of differences within the relevant public and then come to terms with sufficiently equitable ICT strategies to respond to these conditions.

Second, as organizations increase engagement with a relevant public through ICT, they must also develop greater awareness and capacity to manage information identifying individual members of the public. There are serious privacy concerns that need to be factored into the discussion (Taylor et al. 2014), especially as more and more countries implement data privacy laws (Greenleaf 2017). Singh, Gurumurthy, and Chami (chapter 10, this volume) do not suggest that identifying a relevant public necessarily means keeping records of individuals who belong to that public.

However, my observations of CSO practices and interviews with CSO directors indicate that CSOs frequently keep records of who has attended events or participated in their programs. They collect details about individuals, such as their name, telephone number, email address, and gender. Some CSOs have sophisticated information management databases, which can be used to encrypt these details or to search for all records of an individual (so as to remove them if necessary). Yet other CSOs did not have access to such systems. Some CSOs kept participant records on insecure servers, sharing this information across countries, making it difficult to find and remove all records for an individual. Moreover, some CSOs did not inform participants about what would be done with their information and did not ask for written consent to use their real names or photographs in publications.

It is important for CSOs to understand their responsibility to protect the private information of their relevant public. It would be a grave misinterpretation of Singh, Gurumurthy, and Chami's (chapter 10, this volume) open institutional design to increase identification of a relevant public without simultaneously addressing privacy concerns.

Problem 2: Open Institutional Design Will Not Resolve Severe ICT Inequality

Open institutional design focuses primarily on how organizations might practice using technology in a more democratic manner. Singh, Gurumurthy, and Chami (chapter 10, this volume) argue that the advantages of doing so lead to less institutional capture, which may also increase participation, collaboration, and transparency. However, many CSOs still have limited options to acquire and use basic Internet connections and technologies for these purposes. My case study involving Togo's La Colombe, Canada's Crossroads International, and Global Affairs Canada (GAC) also demonstrated how difficult it can be to resolve the problem of severe ICT inequality. Yet Singh, Gurumurthy, and Chami's (chapter 10, this volume) discussion of the problem of institutional capture and the proposed response of a "multiple-shell structure of nested networks" provides good reasons why severe ICT inequality should not be ignored.

To illustrate what I mean by severe ICT inequality, I examine how international cooperation between Togo's La Colombe and Canada's Crossroads International functioned. La Colombe is an organization focused on delivering women's and girls' empowerment programs and providing job skills training at an educational center outside of Togo's capital, Lomé, in the Vo Prefecture, a maritime region in Togo. In the context of practice, almost none of the women participating in La Colombe's education programs had access to technology. When organizing meetings or events, La Colombe staff sent letters by courier and visited rural villages in person. At the La Colombe satellite office in Vo, staff shared a single Internet dongle, which frequently stopped working because the service provider shut down service regularly. Daily electricity cuts also occurred for hours at a time. Furthermore, La Colombe owned desktop computers with expired operating system licenses, and some were dysfunctional because of viruses. Considering that many of the staff worked closely with community members at the education center ten kilometers away or in the villages across the region, most staff members were rarely present in the office to use the computers available.

In this context, it may seem absolutely unrealistic for La Colombe to consider using technology to engage their relevant public, yet the organization relies heavily on partnerships with CSOs, such as Crossroads International, to fund and develop its programs and services. For instance, Crossroads

International sends Canadian volunteers with small budgets to complete pieces of work with La Colombe. Much of their program development is mediated by relationships with international organizations, through which much communication and negotiation happens online. However, severe ICT inequality between La Colombe (in the practice context) and its international partners enabled La Colombe to share information selectively. During my fieldwork, it was clear that La Colombe was not interested in sharing information about its work transparently or involving their relevant public in organizational decision-making in participatory ways. I observed staff members share the progress of their projects with visiting donors differently—sometimes they were honest about what was not working, whereas at other times they made it seem as though all was running smoothly. It was difficult to obtain any documentation about their projects, yet they confirmed having this documentation. Of course, there are many organizational, cultural, and personal reasons why the staff and organization chose not to share or collaborate transparently (Bentley 2017). However, such findings broadly confirm the importance of open institutional design in reducing institutional capture.

In contrast, La Colombe's institutional partners, Crossroads International and GAC, did not view ICT inequality as an issue to resolve. Donors often recognize the power that they hold within development aid relationships, but they often try to minimize the issue rather than reduce these inequalities (Wallace, Bornstein, and Chapman 2006). Crossroads International avoided imposing Canadian managerial procedures on their partners, but this strategy negated the need for partners to develop the administrative capacities to contribute cross-culturally and ultimately increase transparency. ICT inequalities reinforced clear differences between the actors and were associated with disjuncture between the approaches that Crossroads International and La Colombe enacted independently.

In sum, CSOs can certainly engage their relevant publics in transparent, participatory, and collaborative ways without needing or using ICTs. My objection relates to the lack of opportunity the relevant publics have to represent themselves within the nested institutional structure. Almost all the information used to make decisions within the nested institutional hierarchy is filtered by La Colombe. Open institutional design could indeed help these organizations think through how transparency, participation, and collaboration may be fostered across significantly different ICT-resourced contexts. However, it will not resolve these differences unless the implicated actors take responsibility for reducing ICT inequalities.

Problem 3: CSOs Dependent on Donor Funding May Not Have Much Power to Adopt Open Institutional Designs

Many development CSOs, particularly those within African countries, where my research took place, are heavily dependent on donor funds, whether they be from private, CSO, foundation, bilateral, or multilateral sources (USAID 2015). This dependency affects the current institutional hierarchy considerably. As my research focused on relationships between bilateral donors and CSOs, such dependency meant that subsidiarity was influenced by bilateral donor institutions that were extending the *outside rules* to funding recipients, thereby playing a significant role in setting the standards to be achieved and determining what is considered effective functioning. CSOs must also abide by the rules put in place by their own country's legislature. Bilateral donor institutions are likewise influenced by their own sets of *outside rules* put in place by their governments and international agreements (Hyden and Mukandala 1999). Such complexities make it challenging to establish links within a nested hierarchy of institutions such that they all align with open institutional design principles. For instance, although many of the major bilateral donor organizations have laws to guide and protect the use of development aid spending to address poverty in developing countries, there have always been national interests that interfere with how development aid money is being spent. Former colonial ties are a major determinant of foreign aid giving (Alesina and Dollar 2000).

After the World Trade Center towers in the United States were attacked on September 11, 2001 (also known as 9/11), the war on terrorism became a backdrop for development aid spending (Fleck and Kilby 2010; Howell and Lind 2009). Now we are seeing a reemergence of development aid spending that must also deliver on national commercial and private sector interests (Elliott 2017; Mackrael 2014; Star Editorial Board 2018). Clearly, it is debatable how confident we can be in trusting these institutions to set the outside rules for the transparency, participation, and collaboration principles CSOs should adopt.

Furthermore, trends in program funding increasingly seem to favor short-term projects focused on specific development objectives rather than the long-term, unrestricted funding that enables CSOs to prioritize the interests of their relevant publics. During interviews with CSO managers and directors, many interviewees, upon reflection, concluded that long-term unrestricted funding opportunities are changing. In Canada, GAC's Partnership Program has historically provided core funding to a finite set of Canadian organizations. As Brown (2012) explained, changes to the program were implemented in 2009–2010, and since then GAC has once again significantly restructured its funding modalities.

Now there are calls for proposals based on themes or priority countries that either Canadian or international organizations (or both) can apply for. A Canadian CSO director commented that the new funding modalities put organizations in a tough position to decide whether and how to adapt: "We're defending the modalities that we are used to only because we have developed these modalities through lessons learned over the past 10 and even 20 years." Many organizations have to decide whether to build new partnerships, or to refocus their work to respond to funding calls in certain countries or on priority themes, which goes against their prior focus on local capacity building of long-term partnerships to help their partners respond to their relevant publics.

Similarly, in the United Kingdom, three of the four UK interviewees reported that many of the partnership programs were ending or had been drastically reduced. For one director of a CSO in the United Kingdom, this contributed to the organization's decision to shut down: "Working as an NGO, every day, every week, you don't know what tomorrow will hold, juggling continually. The framework grants really give you the room to breathe, securely, otherwise you're constantly firefighting. It's absolutely vital. Given current mania among donors, moving away from that into highly instrumentalist, contractual this for this, away from highly individualized projects into big consortium led programs, it's a nightmare for small NGOs. There's no way we can compete." While it is likely that competition for funds has drastically increased, especially since many southern organizations are now able to compete with their northern counterparts, the reduction of unrestricted funding opportunities was hindering the CSOs' capacity to prioritize open institutional ideals.

Moreover, two of the organizations that had contribution agreements (one was based in Canada and the other was based in the United Kingdom) are less reliant on them now than they were in the 1990s, which they said was because they had diversified their funding sources in the decade that followed. Although these interviewees reported that project-based funding is likely to be swayed by shifting donor priorities, they thought that having a concentrated organizational mission and diversified funding sources enabled their organizations to thrive. Yet, the difference between these two organizations and those more reliant on unrestricted funds was that their programs were oriented toward more technical development approaches than the others, which approached development through capacity building, empowerment, and participation. Community-driven development often engages participants as decision makers, which meant that the CSOs whose members I interviewed wanted to maintain responsiveness to their relevant publics. However, when impact is difficult to measure or does not demonstrate concrete results within a one- or two-year time span, interviewees reported that it was harder to diversify their funding sources.

These findings suggest that for CSOs to adopt open institutional designs, their funding partners must do so as well. While Singh, Gurumurthy, and Chami concentrate on how higher levels of the institutional hierarchy establish outside rules for those lower in the hierarchy to structure transparency, participation, and collaboration, I emphasize that there must also be a strong and unified base of public interest organizations that are capable of advocating for open institutional design across the aid chain—imagine an advocacy network shaped like an upside-down T. For open institutional design to work in this context, CSOs must concentrate their efforts not only on their global policy agendas but also on establishing independent analysis of processes and procedures by actors across aid chains (the upside-down T). This would ensure multiple perspectives on policy-making options and allow different points of view to be recorded and debated. For bilateral donors to seriously consider this option, a social and political movement, rather than merely a new open institutional design, is warranted.

References

Alesina, Alberto, and David Dollar. 2000. "Who Gives Foreign Aid to Whom and Why?" *Journal of Economic Growth* 5 (1): 33–63.

Banks, Nicola, David Hulme, and Michael Edwards. 2015. "NGOs, States, and Donors Revisited: Still Too Close for Comfort?" *World Development* 66:707–718.

Bebbington, Anthony J., Samuel Hickey, and Diana C. Mitlin. 2008. "Introduction: Can NGOs Make a Difference? The Challenge of Development Alternatives." In

Can NGOs Make a Difference?: The Challenge of Development Alternatives, edited by Anthony J. Bebbington, Samuel Hickey, and Diana C. Mitlin, 3–37. London: Zed Books.

Bentley, Caitlin M. 2017. "Bilateral Donors and Civil Society Organisations: Technologies for Learning and Accountability." Doctoral thesis, Royal Holloway University of London.

Brown, Stephen. 2012. "Aid Effectiveness and the Framing of New Canadian Aid Initiatives." In *Struggling for Effectiveness: CIDA and Canadian Foreign Aid*, edited by Stephen Brown, 79–107. Montreal and Kingston: McGill-Queen's University Press.

Desai, Vandana, and Rob Imrie. 1998. "The New Managerialism in Local Governance: North-South Dimensions." *Third World Quarterly* 19 (4): 635–650.

Elliott, Larry. 2017. "Impact of UK Foreign Aid Diluted by Pursuing National Interest, Says IFS." *The Guardian*, May 8. http://www.theguardian.com/business/2017/may/08/foreign-aid-national-interest-ifs-oda-spending.

Eyben, Rosalind. 2013. "Uncovering the Politics of 'Evidence' and 'Results': A Framing Paper for Development Practitioners." Brighton: Institute of Development Studies, April 1, 2013. https://www.ids.ac.uk/publications/uncovering-the-politics-of-evidence-and-results-a-framing-paper-for-development-practitioners/.

Fleck, Robert K., and Christopher Kilby. 2010. "Changing Aid Regimes? U.S. Foreign Aid from the Cold War to the War on Terror." *Journal of Development Economics* 91 (2): 185–197.

Greenleaf, Graham. 2019. "Countries with Data Privacy Laws—by Year 1973–2019 (Tables)." *Privacy Laws & Business International Report* 159:18–21. https://papers.ssrn.com/sol3/papers.cfm?abstract_id=2996139.

Howell, Jude, and Jeremy Lind. 2009. "Changing Donor Policy and Practice in Civil Society in the Post-9/11 Aid Context." *Third World Quarterly* 30 (7): 1279–1296.

Hulme, David, and Michael Edwards, eds. 2013. *NGOs, States and Donors: Too Close for Comfort?* 2nd ed. London: Palgrave Macmillan.

Hyden, Goran, and Rwekaza Mukandala. 1999. "Studying Foreign Aid Organizations: Theory and Concepts." In *Agencies in Foreign Aid: Comparing China, Sweden and the United States in Tanzania*, edited by Goran Hyden and Rwekaza Mukandala, 8–30. London: Palgrave Macmillan.

Lewis, David, and Nazneen Kanji. 2009. *Non-governmental Organizations and Development*. Routledge Perspectives on Development. London: Routledge.

Mackrael, Kim. 2014. "Commercial Motives Driving Canada's Foreign Aid, Documents Reveal." *Globe and Mail*, January 8. https://www.theglobeandmail.com/news

/politics/commercial-interests-taking-focus-in-canadas-aid-to-developing-world /article16240406/.

Star Editorial Board. 2018. "It's Time for Canada to Do Better on Foreign Aid." *The Star*, January 15. https://www.thestar.com/opinion/editorials/2018/01/15/its-time-for -canada-to-do-better-on-foreign-aid.html.

Taylor, Linnet, Josh Cowls, Ralph Schroeder, and Eric T. Meyer. 2014. "Big Data and Positive Change in the Developing World." *Policy & Internet* 6 (4): 418–444.

United States Agency for International Development (USAID). 2015. *2015 CSO Sustainability Index for Sub-Saharan Africa*. Washington, DC: USAID. https://www.usaid .gov/sites/default/files/documents/1866/2015_Africa_CSOSI.pdf.

Wallace, Tina, Lisa Bornstein, and Jennifer Chapman. 2006. *The Aid Chain: Coercion and Commitment in Development NGOs*. London: ITDG Publishing.

Wallace, Tina, and Fenella Porter, with Mark Ralph-Bowman, eds. 2013. *Aid, NGOs and the Realities of Women's Lives: A Perfect Storm*. Rugby: Practical Action Publishing.

14 Conclusion

Matthew L. Smith, Arul Chib, and Caitlin M. Bentley

The impetus for this book was to develop theory that cut across domains of open development. This stemmed from the observation that a lot of "openness" work in the Global South (and globally, for that matter) happens in silos with little to no communication across domains. Open education did not learn from open data, which did not learn from open innovation, and so on, despite similar roots. Siloed practice certainly results partly from institutional influences (academic, funders, journals, etc.) that typically do not incentivize cross-disciplinarity, but it also results from definitional disparities that obscure the similarities that could be the basis for knowledge sharing. Crosscutting theory, the thinking goes, would highlight these areas of similarity and facilitate this cross-disciplinary engagement.

This book's contributors on theory development addressed this challenge by proposing frameworks on various crosscutting themes, such as stewardship, trust, situated learning and identity, understanding inequalities, critical capabilities, and increasing transparency, participation, and collaboration within institutions. However, teams of contributors, composed of scholars from different disciplines in order to encourage innovation, faced challenges in departing from their established disciplinary approaches. In order to investigate these themes in terms of the impact of practices and processes of openness on development, we first needed to foster interdisciplinary integration by establishing a common understanding of open development.

To facilitate this process, contributors met for two days and established a common definition of open development: *the sharing of free, public, networked information and communication resources for positive social transformation.* This characterization mirrored past definitions of open development that focused on defining open or openness, leaving the term *development* underspecified. This approach is reasonable, as there are many opinions on what development is and because the key term that needed defining was *openness* (see Bentley, Chib, and Smith, chapter 1, this volume).

This approach, however, proved to be contentious, as it prioritizes a focus on open processes in international development contexts rather than a more normative notion of what open development is or ought to be. The discussion broadly focused on bridging the interaction between users and the social power structures that shape openness, with debates about approaches prioritizing the user's experience with technology (in a broad sense, technodeterminist) versus those that applied normative evaluation frameworks (broadly speaking, from a critical perspective), encapsulated by Singh, Gurumurthy, and Chami (chapter 10, this volume), who state, "Our starting point is that development is certainly a normative discipline, and open development also needs to be seen as such." A more robust transdisciplinary unity in approaching the issue of open development would require reconciling these perspectives, despite the fuzzy boundaries and messy encounters involved (Chib and Harris 2012).

The chapters in this book bridge these two perspectives by making theoretical connections between open processes and the broader social context, and consequentially with the idea of positive social transformation. In doing so, the arguments presented advance our theoretical understanding of open development. They do so in three clear ways. First, we acknowledge that a simple process orientation is too narrow a definitional focus, which makes it difficult to connect openness to broader social outcomes. Second, openness (or open processes) can be aptly thought of as elements of social relationships with inherent power differentials, and open development is a normative perspective on how these open processes might alter these unequal relationships. In this way, the chapters start to strengthen our theoretical connection between open processes and development outcomes. Third, such thinking necessitates a critical perspective with a focus on the power dynamics in these relationships from design to evaluation of open initiatives.

This chapter spells out these insights in more detail. It starts with some thoughts on the term *positive transformation* across the chapters. We then discuss how openness contributes to a process of positive social transformation by comparing and contrasting the theoretical approaches presented alongside the empirical reflections.

Understanding Positive Social Transformation

Before the outset of this project, we invited a range of experts to debate the key issue of siloed research in the area of open development. One consequence of the failure to disentangle the crosscutting features and conditions of digitally enabled openness was being able to translate this understanding into policies and practices. We chose to allow the process to be emergent by not taking a normative position on what positive social transformation would be. Researchers were given the flexibility to define it themselves. Nor was there explicit direction in terms of the focus of the theory. So, the first question is, is there anything we can learn about what positive social transformation might be from a synthesis of the chapters in this volume?

On the surface, the book's chapters on theory sliced open development in a variety of ways, highlighting different elements of open development. In chapter 2, Reilly and Alperin examine stewardship of open data in relation to commons governance. In chapter 3, Rao et al. examine the impact of potential relationships of trust between different actors in open systems. In chapter 4, Chaudhuri, Srinivasan, and Hoysala explore how instrumental, substantive, and situated informal learning of new literacies are influential components of the processes that make up these relationships. In chapter 8, Dearden, Walton, and Densmore examine how reading and writing relationships, via literacy events, influence the appropriation, engagement, and outcomes when people weave new open technologies into their daily activities. In chapter 9, Zheng and Stahl offer the concept of critical capabilities as an evaluation space for open development processes as well as an approach to the design of open initiatives. Finally, in chapter 10, Singh, Gurumurthy, and Chami provide a specific interpretation of open development as the use of open processes to create truly open institutions that operate in the public interest.

Interestingly, despite the different emphases of these chapters, a common theme emerges. Explicitly or implicitly, all the chapters characterize openness as part of a larger (set of) social relationship(s) (see table 14.1). Furthermore, open processes are one potential way to bring about a transformation in that relationship. For the authors in this book, this is social transformation.

The notion of social transformation as an alteration of power relationships is made explicit in most of the chapters on theory. In chapter 2, Reilly

Table 14.1
Social transformation as changing social relations in open initiatives

Topic	Quotations on Transformation in This Volume
Stewardship	"Engagement, however, implies motivated and reflexive contributions to a jointly produced and therefore evolutionary context. It recognizes the dissolution of the boundary between user and producer in the management of the resources, *as well as the shifting balance of costs and benefits between different user groups* [our emphasis], which might include a wide range of actors. … As a result, engagement is said to be *transformative* [emphasis original] in nature" (Reilly and Alperin, chapter 2).
Trust	"Our model offers *a relational view of trust* [our emphasis] that can be used to overcome trust issues hindering the contribution of openness toward a process of positive social transformation" (Rao et al., chapter 3).
Learning as participation	Examining both individual and social transformation,"How people learn [emphasis original] to make sense of and cope with the contextualized changes induced by participating in open practices is itself a social transformation" (Chaudhuri, Srinivasan, and Hoysala, chapter 4). "Substantive aspects of learning will always appear *in relation to* [emphasis original] increasing participation in a CoP [community of practice], which is inherently governed by power relations in context" (Chaudhuri, Srinivasan, and Hoysala, chapter 4).
Divergent outcomes	"We argue that the situated informal learning of new literacies is a critical element that will be required if open services are to contribute to positive social transformation. Material inequalities in technical infrastructure and tools will constrain people's ability to convert access to open services into outcomes for themselves. *Social inequalities will restrict people's ability to adjust their activities to take advantage of and contribute to new services* [our emphasis]. Social rules, social connections, and the command of literacy practices will constrain the relative ability of participants to engage with and learn to shape these spaces" (Dearden, Walton, and Densmore, chapter 8).
Critical capability approach	"The CCA can be used to *assess the motivations driving open development initiatives and their social consequences* [our emphasis]. The CCA consists of the following four principles: 1. The principle of human-centered development; 2. The principle of human diversity; 3. The principle of protecting human agency; 4. The principle of democratic discourses" (Zheng and Stahl, chapter 9).
Open institutions	"This definition places the customary constituents of *openness* [emphasis original]—transparency, participation, and collaboration—in a situated institutional setting, *as a set of social relationships among specific social actors* [our emphasis]" (Singh, Gurumurthy, and Chami, chapter 10).

and Alperin write about *engagement* in openness initiatives and contrast it with targeted participation. For these authors, targeted participation tends to be *transactional*, whereas engagement "implies motivated and reflexive contributions" and "recognizes the dissolution of the boundary between user and producer in the management of the resource, as well as the shifting balance of costs and benefits between different user groups." In chapter 3, Rao et al. begin by taking a "relational view of trust that can be used to overcome trust issues hindering the contribution of openness toward a process of positive social transformation." Depending on the open development initiative, there are many potential trust relationships between different stakeholders from quite direct and personal to indirect and mediated by technology. Therefore, social transformation involves the creation of new social relationships or the changing of existing ones, within which trust is a key factor.

In chapter 4, Chaudhuri, Srinivasan, and Hoysala talk of positive social transformation as being "*in relation to* increasing participation in a CoP [community of practice]," noting the sociocultural contexts of power within which open initiatives occur. These authors point to the differential open development outcomes for different people, depending on their inclusion or exclusion as a function of their social positions. In chapter 10, Singh, Gurumurthy, and Chami take an active stance on the normative view of open development, focusing on the participation of, and impact on, marginalized populations. These authors develop an institutional perspective, questioning the broader social impact of open development in terms of power relationships between internal institutional actors and the relevant public interest.

Two other chapters on theory also take a relational view, but more implicitly. These chapters talk about how social transformation is not the necessary outcome of open processes but rather is a function of the larger context of social inequalities that shape the potential for positive social transformation. In chapter 8, Dearden, Walton, and Densmore differentiate between reading and writing relationships in informal learning, positing that this can lead to people "debating and shaping their own futures and challenging existing power relations." These authors, however, caution that these potential positive outcomes face constraints, arguing that existing social positions and peer relationships influence participation and interaction in open services. They make a case for the need to focus an intervention "toward these material and social inequalities" to achieve social transformation. In chapter 9, Zheng and Stahl argue that the notion of capabilities focuses on conversion factors that are both individual and stem from an individual (or group) positionality with social structures. Consequently, capabilities are relational (Smith and Seward 2009), and transformation requires some restructuring of those relationships.

These chapters, plus the empirical reflections, suggest that we can understand the outcomes of open development on at least two levels. The first level is of direct, first-order outcomes, which can be seen through the lens of open processes. These are typically the instrumental outcomes wherein an individual or family's life may be improved, for example through cost savings, free access to educational content, or sharing of agricultural data, as described in chapter 11 by Gamage, Rajapakse, and Galpaya in the Sri Lankan case. It is not surprising that the bulk of impact research looks at these types of changes (see also Bentley and Chib 2016; Bentley, Chib, and Poveda 2019). The second level of change may include these types of improvements, but takes a broader perspective to examine the transformation of the underlying social relationships. For the authors in this book, it is fair to say that it is only this second level of change that constitutes positive social transformation.

Connecting Openness to Social Transformation

We do not think it coincidental that while grappling with the idea of cross-cutting theory for open development (and not just openness) the authors coalesced around notions of social transformation and social relations. Of course, development is at its core a social change process, inherently one that involves (unequal) distributions of resources (power) in a network of social relationships.

Smith and Seward (2017, 2020), describing openness as a "social praxis," delineated four open practices that informed many chapters in this volume: peer production, crowdsourcing, sharing, and reuse (consumption). While Smith and Seward did not make it explicit, it is easy to see how all these are predicated on some form of social relationship. Peer production and crowdsourcing, of course, clearly are new organizational forms with new roles and relationships that enable production of content. Sharing and reuse, however, are also a relationship in and of themselves: "Sharing is only realized when what is shared is consumed; and 'consumption practices' are only open when they involve the consumption of something that has been shared" (Smith and Seward 2017, n.p.).

What this book's contributors suggest, then, is that we should understand open practices as a constituent activity in constructing and maintaining social relationships. A relational model brings in process, people, and structural relations (which includes people's relations to things). As Dearden, Walton, and Densmore (chapter 8, this volume) write, "relationships ... includes both people and the materialities that allow them to relate to one another." We see in these chapters that these relationships can be multidimensional, including communities, stakeholders, technologists, and practitioners, some of whom are proximal to each other whereas others are not. Interestingly, and perhaps not surprisingly given the notions of crowdsourcing and peer production, several chapters bring in the idea of communities of practice.

Conceptualizing open processes as part of a social relationship allows the beginning of developing a clearer theoretical connection with social change processes that make up development (or positive social transformation). In doing so, this theoretical approach provides insight into thinking, design, and evaluation of open development initiatives going forward.

The first insight is that supply-side approaches (such as those described in the arterial school by Reilly and Alperin in chapter 2) need to look beyond the instrumental effects of an open intervention to find evidence of positive social transformation. This is not a new critique and can be seen in a lot of research on openness in development (Bentley, Chib, and Poveda 2018). For example, there is a worry that open processes, as they currently tend to be conceived and implemented, will contribute to social inequalities, particularly given the context of inequality (Smith and Seward 2020). In the case of open data, when an initiative is mediated by arterial school principles, actors presume that facilitating open data arteries will enable their effective use. This is because the arterial school assumes fundamentally that opportunity is equal (Dearden, Walton, and Densmore, chapter 8, this volume). In other words, everyone has the same capacity and opportunity to take advantage of what is shared.

However, Sen (1992) powerfully critiques this notion of equality, stating that the evaluation space for equality should be that of capabilities to do things one has reason to value rather than having access to resources only—especially given that different people have different *conversion factors* to convert the resources into functionings (or outcomes). For example, in chapter 7, Kendall and Dasgupta illustrate the example of Moumita, who was selected to collect weather data and update the public blackboard in her village. She was able to do so because she was the daughter of a respected person, was literate, and had completed a high school education. On the other hand, unlike men, it was more common for women to be excluded from viewing blackboards on a daily basis because the blackboards were positioned in the village centers.

Both empirical reflections examining open government data initiatives illustrate the need to look, from an evaluation perspective, beyond the first-order instrumental impacts of an intervention. In chapter 12, Moshi and Shao contrast the evaluation approach of Tanzania's former open government initiative with the critical capability approach to evaluating open government data. The critical capability approach (Zheng and Stahl, chapter 9, this volume) immediately turned up issues beyond first-order instrumental impacts (in this case, increased transparency through the publication of data sets). Likewise, Mungai and Van Belle (chapter 5, this volume) found a similar tendency for data fellows within Kenya's open government data initiative to focus on supply-side open data processes rather than addressing meaningful use requirements of citizens. In this way, as Singh, Gurumurthy, and Chami (chapter 10, this volume) point out, a process orientation can easily lead to open washing, where a narrow focus on acts of openness can be used to gloss over activities that might worsen inequalities.

Of course, this is not an entirely new insight in the development field, which has talked about failings of highly technocratic and depoliticized development interventions (Ferguson 1990). However, the pattern does seem to repeat whenever a new and promising technology comes along. In the open development space, this was seen in the plethora of approaches that prioritized opening information for its own sake, focusing on the technical details of opening content, and assuming that use would follow. For example, in the open data initiative in Tanzania reported by Moshi and Shao (chapter 12, this volume), the open data was a standalone resource that had little influence on the priorities or policies of the Ministry of Education. In chapter 2, Reilly and Alperin call this the "arterial school" and suggest it applies beyond open development. For example, many "open" activities are just simply sharing: open access scholarly articles, open educational resources, and open research data are typically just made openly available, without any further activity. Furthermore, where it is used in lower- and lower-middle-income contexts, it appears to be mostly limited use as is, rather than aiming to modify, adapt, and reshare (Smith and Seward 2020). However, research on open development has demonstrated that in many instances a deeper process of engagement that promotes and facilitates use is necessary, such as, for example, when intermediaries take and/or modify the shared content and integrate it into wider systems (Seward 2020). For example, open data for development interventions are shifting to take an ecosystem approach, moving beyond just supply-side issues to understanding and engaging with an ecosystem of actors that ultimately contribute to the *use* of the data (Davies et al. 2019; van Schalkwyk and Cañares 2020).

Yet, as we saw in chapters 6 and 11, the power relations within a system often dictate the roles and positions that actors take within open initiatives. In chapter 11, Gamage, Rajapakse, and Galpaya demonstrated how multiple activity systems often coexist and compete within open initiatives. The role of these systems within the project was critical for mediating power differentials between the agricultural department actors who were responsible for creating the agricultural information sharing app and the farmers who needed to use the information and contribute back. Similarly, in chapter 6, Sadoway and Shekhar argued that citizen perspectives really must be prioritized in open urban service initiatives and that, without direct (and equal) relationships between citizens and service providers, trust will degrade within open systems.

These empirical reflections take value positions in terms of how open processes *should* contribute to social transformation. For example, in chapter 11, Gamage, Rajapakse, and Galpaya state that "*writing rights* implies that open processes enable [Sri Lankan] farmers to have a voice, to reuse and repurpose the information, or to contribute through their own experience and knowledge." However, is that *necessary* for social transformation? Open processes engage with digital content in three main phases: creation, use, and adaptation (Trotter and Hodgkinson-Williams 2020). According to Wiley (2014), while *use as is* is important, the real transformational value comes from being involved in either the creation or adaptation. Yet, the answer to the preceding question is both yes and no. We argue that both the pragmatic and coevolutionary approaches to open development have consistently shown how researchers and practitioners can both identify and improve social transformation outcomes (or lack thereof). Both types of theoretical lenses are valuable in working through the types and magnitude of social transformation outcomes, and they are not necessarily dependent on the type of openness enacted.

To illustrate this, the pragmatic approaches in part I of this volume provided a language for identifying social transformation regardless of the types of social relationships in play or the type of open activity (use, adaptation, and modification). The freedoms (e.g., to create, curate, adapt, and critique) that open processes can allow may result in an important shift in perceptions of expertise, and the voices of people who were otherwise left out may be included. Chaudhuri, Srinivasan, and Hoysala (chapter 4, this volume) posited that people who move toward increasing participation in communities of practice generate new relationships and identities in open initiatives, which was confirmed by Kendall and Dasgupta (chapter 7, this volume) for certain individuals. This is social transformation.

Reading rights (i.e., as-is use) can lead to social transformation as well. In chapter 3, Rao et al.'s model offers a way to identify various actors in an open system to examine how trust affects a user's engagement. When people around the world take a massive open online course (MOOC), they will not, for the most part, alter their relationship with the institution that created that MOOC. Yet, the ability to get an education that is trusted by employers can transform someone's life, even if they were not an active creator of the educational content shared with them. Similarly, in chapter 2, Reilly and Alperin's stewardship approach to open data highlights how there are multiple ways that actors contribute social value (see also van Schalkwyk and Cañares 2020). Although Mungai and Van Belle (chapter 5, this volume) would like to see a greater focus on meaningful use of open government data in Kenya, the stewardship regimes of the data fellows within the Kenyan government have transformed certain internal government functions, which may enable the government to institutionalize improved services for its citizens. In these cases, the pragmatic approaches highlight how multiple types of open processes can lead to social transformation but do not necessarily change social relationships between the creators and users of open resources. The pragmatic approaches enable the flexibility to focus on social transformation outcomes that are perhaps the most important in a particular context at a specific moment in time.

Moreover, openness as *trust*, as Rao et al. describe in chapter 3, suggesting credibility and confidence in an open system, partly depends on the relationships between the various stakeholders and actors involved, including users and developers. Sadoway and Shekhar (chapter 6) describe a top-down smart city initiative in Chennai with limited public engagement or feedback forums. For these authors, the contextual complexities of the local environment require that trust be examined within an existing web of multilevel power dynamics. They argue that traditional notions of the beneficiary public view them as *external* agents in a *sponsored* system, which fails to account for the public as (pro) active citizens and comanagers of open information. They suggest the building of *urban trust networks* as a first step toward provision of basic public services for the marginalized poor.

Indeed, we argued that the authors of the chapters in part II tend to favor participatory and democratic processes leading to significant evolution in relationships between creators and users, as well as how relationships and principles are embedded into institutions. Dearden, Walton, and Densmore's (chapter 8, this volume) stance is that actors do need to have *writing rights*, or the capacity to alter or modify open resources. Yet, for the most part, these authors believe strongly that the *context* in which actors contribute (i.e., based on participatory, democratic, human-centered principles) matters more than whether they contribute or not. Sometimes, however, openness as *transparency*, as Singh, Gurumurthy, and Chami discuss in chapter 10, might change that relationship even if there is only use as is. For transparency to result in a change in the relationship between the institution sharing information and the "relevant public" that is receiving that information, it must lead to a relationship whereby the transparency enables accountability. This requires that the information shared be appropriate for external accountability and that there be legitimate mechanisms through which the institution can be held accountable. It is not surprising, for example, that the most effective civic tech implementations to generate government responsiveness were those that were already connected with or supported by government (Bonina and Scrollini 2020; Peixoto and Fox 2016). This connection creates a route by which the information flow can inform government behavior and be responsive.

A final thought concerns reflexivity, not merely for the editors and contributors to this volume but for the field of open development research as a whole. Having established social relations as a key component to enable the potential of sociostructural transformation via open development to be achieved, we can question the normative value of openness as a means to transform social position in development via research. As remarked on previously, much of the literature focuses on first-order instrumental impacts, largely at the individual level. Bentley and Chib (2016), in a review of open development in low- and middle-income countries, found scarce evidence to support the view that research is concerned with the perspectives of poor and marginalized people, providing support for earlier critiques on the distribution of benefits from openness (Buskens 2011; Gurstein 2010). In this volume, too, Mungai and Van Belle (chapter 5) draw our attention to the actors absent from their analysis of stewardship regimes in Kenya's open data initiative, namely those mired in poverty. Furthermore, acknowledging the role of participation in social transformation, a mere 2 percent of 269 studies on open development incorporated participatory designs (Bentley and Chib 2016). It is of concern that research in open development relies more on self-identification as a field than on application of the principles we advocate.

In sum, this quick synthesis of the chapters highlights a way to make the theoretical connection between open processes and social transformation, defined as a change in unequal power dynamics of social relationships. To encapsulate our assessment influenced by the contributors, we need to think of open development beyond process definitions by adopting critical considerations of social positionality and relational power differentials. Open development then takes on normative perspectives on how open processes alter unequal social relationships, particularly for marginalized communities. Doing so allows us to be more systematic and intentional about going beyond a focus on first-order effects, which was so commonly the focus in early open development implementation and research.

Conclusion

One contribution of this volume is to extend the notion of openness as social praxis to incorporate social relations and positionality. Openness practices happen in communities, sometimes of disparate stakeholders, sometimes online and impersonal, and at other times in the very local meaning of community.

It should be no surprise, then, that the authors of this book agreed that a key theme was the need for a critical perspective. Once one accepts the existence of social structures, their examination necessarily implies a critique.

We therefore suggest the need to question and understand historically and culturally situated contexts of marginalized individuals and the politically motivated, albeit unintentional, beliefs of proponents of open development. Bentley, Chib, and Poveda (2018) argue that we should consider both the empowering and disempowering effects of openness, critically examining the intention of researchers as actors in the process of structural transformation. There is no doubt that openness is a powerful tool that can change relational and structural elements of society. If open development is to contribute to positive social transformation, we encourage first understanding and identifying the underlying sociostructural context in terms of unequal power relationships and social positionality before designing and sharing participatory and transparent networked solutions.

References

Bentley, Caitlin M., and Arul Chib. 2016. "The Impact of Open Development Initiatives in Lower- and Middle-Income Countries: A Review of the Literature." *Electronic Journal of Information Systems in Developing Countries* 74 (1): 1–20. https://onlinelibrary.wiley.com/doi/10.1002/j.1681-4835.2016.tb00540.x.

Bentley, Caitlin M., Arul Chib, and Sammia C. Poveda. 2018. "Exploring Capability and Accountability Outcomes of Open Development for the Poor and Marginalized: An Analysis of Select Literature." *Journal of Community Informatics* 13 (3): 98–129. http://ci-journal.org/index.php/ciej/article/view/1423.

Bentley, Caitlin M., Arul Chib, and Sammia C. Poveda. 2019. "A Critical Narrative Approach to Openness: The Impact of Open Development on Structural Transformation." *Information Systems Journal* 13:787–810.

Bonina, Carla M., and Fabrizio Scrollini. 2020. "Governing Open Health Data in Latin America." In *Making Open Development Inclusive: Lessons from IDRC Research*, edited by Matthew L. Smith and Ruhiya Kristine Seward, 291–316. Cambridge, MA: MIT Press; Ottawa: IDRC. https://idl-bnc-idrc.dspacedirect.org/bitstream/handle/10625/59418/IDL-59418.pdf?sequence=2&isAllowed=y.

Buskens, Ineke. 2011. "The Importance of Intent: Reflecting on Open Development for Women's Empowerment." *Information Technologies & International Development* 7 (1): 71–76.

Chib, Arul, and Roger Harris. 2012. *Linking Research to Practice: Strengthening ICT for Development Research Capacity in Asia*. Singapore: Institute of Southeast Asian Studies; Ottawa: IDRC. https://idl-bnc-idrc.dspacedirect.org/bitstream/handle/10625/49232/IDL-49232.pdf?sequence=1&isAllowed=y.

Davies, Tim, Stephan B. Walker, Mor Rubinstein, and Fernando Perini, eds. 2019. *The State of Open Data: Histories and Horizons*. Cape Town: African Minds; Ottawa: IDRC. https://www.idrc.ca/en/book/state-open-data-histories-and-horizons.

Ferguson, J. 1990. *The Anti-politics Machine: "Development," Depoliticization, and Bureaucratic Power in Lesotho*. Cambridge: Cambridge University Press.

Gurstein, Michael B. 2011. "Open Data: Empowering the Empowered or Effective Data Use for Everyone?" *First Monday* 16 (2). http://journals.uic.edu/ojs/index.php/fm/article/view/3316.

Peixoto, Tiago, and Jonathan Fox. 2016. *When Does ICT-Enabled Citizen Voice Lead to Government Responsiveness?* Washington, DC: World Bank.

Sen, A. 1992. *Inequality Re-examined*. Oxford: Clarendon Press.

Seward, Ruhiya Kristine. 2020. "Conclusion: Understanding the Inclusive Potential of Open Development." In *Making Open Development Inclusive: Lessons from IDRC Research*, edited by Matthew L. Smith and Ruhiya Kristine Seward, 431–445. Cambridge, MA: MIT Press; Ottawa: IDRC. https://idl-bnc-idrc.dspacedirect.org/bitstream/handle/10625/59418/IDL-59418.pdf?sequence=2&isAllowed=y.

Smith, Matthew L., and Carolina Seward. 2009. "The Relational Ontology of Amartya Sen's Capability Approach: Incorporating Social and Individual Causes." *Journal of Human Development and Capabilities* 10 (2): 213–235.

Smith, Matthew L., and Ruhiya Seward. 2017. "Openness as Social Praxis." *First Monday* 22 (4). https://doi.org/10.5210/fm.v22i4.7073.

Smith, Matthew L., and Ruhiya Kristine Seward, eds. 2020. *Making Open Development Inclusive: Lessons from IDRC Research*. Cambridge, MA: MIT Press; Ottawa: IDRC. https://idl-bnc-idrc.dspacedirect.org/bitstream/handle/10625/59418/IDL-59418.pdf?sequence=2&isAllowed=y.

Trotter, Henry, and Cheryl Hodgkinson-Williams. 2020. "Open Educational Resources and Practices in the Global South: Degrees of Social Inclusion." In *Making Open Development Inclusive: Lessons from IDRC Research*, edited by Matthew L. Smith and Ruhiya Kristine Seward, 317–356. Cambridge, MA: MIT Press; Ottawa: IDRC. https://idl-bnc-idrc.dspacedirect.org/bitstream/handle/10625/59418/IDL-59418.pdf?sequence=2&isAllowed=y.

van Schalkwyk, François, and Michael Cañares. 2020. "Open Government Data for Inclusive Development." In *Making Open Development Inclusive: Lessons from IDRC Research*, edited by Matthew L. Smith and Ruhiya Kristine Seward, 251–290. Cambridge, MA: MIT Press; Ottawa: IDRC. https://idl-bnc-idrc.dspacedirect.org/bitstream/handle/10625/59418/IDL-59418.pdf?sequence=2&isAllowed=y.

Wiley, David. 2014. "The Access Compromise and the 5th R." *Iterating toward Openness* (blog), March 6. https://opencontent.org/blog/archives/3221.

Contributors

Juan Pablo Alperin is an assistant professor at the School of Publishing at Simon Fraser University, Canada, the associate director of research for the Public Knowledge Project, and the co-director of the Scholarly Communications Lab. He is a multidisciplinary scholar, with a bachelors in computer science, a masters in geography, and a PhD in education, who believes that research, especially when it is made freely available, has the potential to make meaningful and direct contributions to society.

Caitlin M. Bentley is a lecturer at the Information School of the University of Sheffield, United Kingdom (UK), and an honorary fellow at the 3A Institute, Australian National University. Her research investigates how AI-enabled cyberphysical systems can be safely, ethically, and sustainably scaled. Specifically, how these systems can be designed and regulated inclusively and governed democratically. She also focuses on ICTs in developing contexts as well as the impacts of AI-enabled cyberphysical systems on marginalized actors. From 2015 to 2017, she worked for the Singapore Internet Research Centre at Nanyang Technological University, pursuing her interests in open development. She holds a PhD in human geography from Royal Holloway University of London, UK, an MA in educational technology from Concordia University, Canada, and a BA in computer science from McGill University, Canada. She is a fellow of the UK's Software Sustainability Institute.

Nandini Chami is deputy director at IT for Change based in Bengaluru, India. Her work focuses on research and advocacy at the intersections of digital policy, development justice, and gender equality. She is part of the core team working on policy advocacy for data justice, inclusive digital trade, and gender perspectives in the debate on ICTs as means of implementation of the SDGs. She also provides strategic support to IT for Change's field center Prakriye in its training programs for women's rights groups on adopting digital tools in their field practice and critical 'education for empowerment' for rural adolescent girls. She has a master's degree in urban and rural community development from the Tata Institute of Social Sciences, Mumbai, India.

Bidisha Chaudhuri holds a PhD from the South Asia Institute, Heidelberg University, Germany and is an assistant professor at the International Institute of Information

Technology-Bangalore (IIIT-B), India. She is the author of the book *E-governance in India: Interlocking Politics, Technology and Culture* (2014, Routledge). Her research interests include e-governance, public policy reform, ICT for development, gender and development, and South Asian politics.

Arul Chib is associate professor at the Wee Kim Wee School of Communication and Information and former director of the Singapore Internet Research Centre at Nanyang Technological University. He investigates the impact of mobile phones in health care (mHealth) and in transnational migration issues, and is particularly interested in intersections of marginalization. He won the 2011 Prosper.NET-Scopus Award for the use of ICTs for sustainable development, accompanied by a fellowship from the Alexander von Humboldt Foundation, one of the highest honors in the European tradition. He has been awarded fellowships at Ludwig Maximilians University and the University of Southern California. He has been the principal investigator and coordinator of the SIRCA (Strengthening Information Society Research Capacity Alliance) programs since 2008, mentoring emerging researchers from Africa, Asia, and Latin America. His research in twelve countries has been profiled in the media ranging from the *United Nations Chronicle* to the Singaporean press. He currently serves on the editorial boards of *Journal of Computer-Mediated Communication, Annals of the International Communication Association*, and *Mobile Media and Communication*, and the publication committee of the International Communication Association.

Purnabha Dasgupta is currently the action research coordinator for the nonprofit Development Research Communication and Services Centre (DRCSC), working on sustainable agricultural development, livelihoods, climate change adaptation, and other issues. Prior to this, he worked for three years in the Department of Science and Technology, Government of India Project on Sustainable Livelihood Development, on scientific management of natural resources. He has submitted his PhD thesis on sustainability assessment of integrated farming systems to Ramakrishna Mission Vivekananda University.

Andy Dearden is professor of interactive systems design at Sheffield Hallam University, where his research deals with participatory methods for designing and using interactive computer and communications systems to support social and economic development for people and communities. He has published widely on how effective design practices, specialist design skills, and good design ideas can be shared to allow people who are not specialist designers to devise workable and appropriate systems. His work is particularly concerned with enabling people and groups who may have limited resources, or limited experience with technology, to shape systems for themselves.

Melissa Densmore is a senior lecturer in the Department of Computer Science at the University of Cape Town (UCT). Prior to joining UCT in August 2014, she completed postdoctoral work at Microsoft Research in Bangalore, India, as part of the Technology for Emerging Markets group, where she has been conducting a trial comparing the effectiveness of community health workers using interactive mobile health education materials to that of health workers using paper flipbooks. Her other work includes a delay-tolerant teleconsultation system for doctors in Ghana and contributions to infrastructure enabling village health centers to consult with doctors at the Aravind Eye Hospitals. Her research interests include human-computer interaction for development (HCI4D), mobile health, and last-mile networking.

Helani Galpaya is chief executive officer of LIRNEasia, a pro-poor, pro-market think tank catalyzing policy change through research. She researches and engages in public discourse on issues related to policy and regulatory barriers in Internet access, human rights online, data governance, and inclusive markets. She serves on the board of editors of *Information Technologies & International Development* (ITID), is on the board of directors of the Global Partnership for Sustainable Development Data (GPSDD), and a member of the multistakeholder advisory group (MAG) of the UN Internet Governance Forum. She is on the advisory board of the Harnessing and Sharing Economy for Local Development initiative of the Centre for Implementation of Public Policies promoting Equity and Growth (CIPPEC), Argentina, and was an advisor to the UN Broadband Commission's Working Group on Bridging the Gender Divide.

Piyumi Gamage's research interests are the use of ICTs in agriculture and supply chain development and studies related to sustainability and climate change. She holds a BS in agriculture (specializing in agricultural economics and extension) from the Rajarata University of Sri Lanka and graduateship in chemistry at the Institute of Chemistry Ceylon. She is a member of the Agricultural and Applied Economics Association and the Sri Lanka Economic Association, an affiliate member of the Chartered Institute of Marketing, United Kingdom, and an associate member of the Institute of Chemistry, Ceylon. She is currently completing her master's degree in agricultural economics at the Postgraduate Institute of Agriculture, University of Peradeniya.

Anita Gurumurthy is a founding member and executive director of IT for Change, an India-based nongovernmental organization that works at the intersection of development and digital technologies. The organizational vision on social justice in the network society draws on Southern critiques of mainstream development, and its key strategy is to create and work through trust-based coalitions and horizontal alliances. Through her work at IT for Change, Anita has attempted to promote conversations between theory and practice. In addition to research responsibilities at IT for Change, she also leads the work of the organization's field resource center, which works with grassroots communities on "technology for social change" models. Equity and community ownership, focusing particularly on socially marginalized women, are the cornerstones of such model building.

Onkar Hoysala is a principal software engineer with experience developing a range of products: simulation tools, SaaS software, and policy research software. He is also trained as a qualitative researcher. He has a strong background designing and developing enterprise grade backend technologies, from the idea-stage to production. Onkar has a master of science (MS) by research in information technology from the International Institute of Information Technology, Bangalore.

Linus Kendall is a PhD candidate at Sheffield Hallam University working on the use of information and communication technologies in development (ICTD). His present research is focused on design of agricultural knowledge management systems for smallholder and subsistence farmers in West Bengal, India. Through this work, he is exploring how ICT designers and designs can engage with questions of ecological sustainability in socioeconomic development.

Rich Ling holds a PhD from the University of Colorado and is currently the Shaw Foundation Professor of Media Technology, Wee Kim Wee School of Communication and Information, Nanyang Technological University, Singapore. He has focused his work on the social consequences of mobile communication. He was a professor at the IT University of Copenhagen, where he assisted in managing the Department of Management, and he works at Telenor, near Oslo, Norway. In 2005, he was the Pohs Visiting Professor of Communication Studies at the University of Michigan in Ann Arbor, where he has an adjunct position. He is the author of the book *Taken for Grantedness* (2012, MIT Press), which was the subject of a complementary review in the journal *Science*.

Goodiel C. Moshi is a lecturer at the University of Dodoma, Tanzania. He earned his BS in information and communication technologies management at Mzumbe University. He obtained his master's and PhD in ICT regulation, policies, and utilization from Waseda University in Japan. His research focuses on ICT policies and utilization in networked societies, particularly in developing countries. His current research interests include ICT infrastructure and utilization, open data, and mobile money in Tanzania and Africa at large.

Paul Mungai is a software developer at the eLearning Support & Innovation Unit (eLSI), University of the Witwatersrand, South Africa. He concentrates on developing and maintaining e-learning web/mobile solutions and carrying out research in ICT4D and ICT4Ed. He also worked with two other leading institutions, the University of Nairobi (2004–2009) and the University of the Western Cape, South Africa (2009–2010). He holds a bachelor of business and information technology (BBIT) degree from Strathmore University and a master of philosophy degree, specializing in information and communication technologies, from the University of Cape Town—Centre for Educational Technology. He is currently pursuing his PhD at the University of Cape Town—Centre for Educational Technology.

Priya Parekh is a digital marketing professional with a background in agency side roles, having managed digital campaigns for the leading skin care and hair care brands—Garnier Skin Naturals and Garnier Fructis (L'Oreal India). She has a master's degree in mass communication from Nanyang Technological University and has worked as a Digital Planner at OMD Worldwide in Singapore.

Chiranthi Rajapakse is a senior researcher at LIRNEasia (http://lirneasia.net) currently working on a project investigating the use of ICTs in agriculture and supply chain development in Sri Lanka. In the past, she has been involved in LIRNEasia research looking at the use of mobile phones among microentrepreneurs in Myanmar and how gender affects mobile phone usage. She has also worked as a feature writer in the print media and holds a bachelor of laws (LLB) degree from the University of London (International Programs) and a degree in dentistry from the University of Peradeniya.

Anuradha Rao is a Singapore-based consultant, educator, trainer, and researcher on cybercrime, cyber-security/safety awareness, and other pertinent issues related to technology and society, politics, and public policy. She is passionate about spreading critical awareness of the perils and promises of new media and new information and communication technologies (ICTs). Anuradha is the founder of CyberCognizanz, a Singapore-based training and communications company that focuses on cyber-safety and cybersecurity awareness. Anuradha has a PhD in communications and new media from the National University of Singapore and an MA in political science from Jawaharlal Nehru University, India.

Katherine M. A. Reilly is director of the MA in Global Communication and associate professor in the School of Communication at Simon Fraser University, Canada. Her research addresses international and development communication, with a particular focus on Latin America. She is co-editor (with Matthew Smith) of *Open Development: Networked Innovations in International Development* (MIT Press and IDRC, 2013) and co-editor (with Anil Hira) of a special edition of the *Journal of Developing Societies* entitled "The Emergence of the Sharing Economy: Implications for Development." She is currently the principal investigator of a comparative research project looking at citizen data in five Latin American countries funded by the International Development Research Centre (IDRC) and the Social Science and Humanities Research Council (SSHRC).

David Sadoway works at the Department of Geography and the Environment at Kwantlen Polytechnic University and was formerly a research fellow with HSS-Sociology at Nanyang Technological University. His current research involves the sociopsychological impact of noise and vibration in high-density neighborhoods. He has a PhD in urban planning and design from the University of Hong Kong (2013) and served as a postdoctoral fellow at Concordia University (Montreal) (2012–2014), where he studied the politics of Indian urban infrastructure in Delhi and Bangalore. He has been a visiting scholar at the Technical University of Darmstadt's Topology of Technology (2013), the National Institute of Urban Affairs, New Delhi (2013), and Academia Sinica's Center for Asia-Pacific Area Studies, Taipei (2008). His research interests include Asian urbanism, civic environmentalism, urban infrastructure and technologies, community informatics, and enclave urbanism. He has worked in the

United Nations system, government, the nonprofit sector, and with urban planning consultants in Toronto and Vancouver.

Deo Shao is a computer systems developer and assistant lecturer in the Faculty of Computer Science, University of Dodoma. He is currently working on projects that attempt to leverage open data for the public good. The first project uses open data in the health sector to improve health service competitiveness and enable the public to make decisions based on quality of health service data. The other project explores the design, process, and structure of open government initiatives in Tanzania. He is pursuing a PhD at the University of Dodoma which focuses on open government data infrastructure in Tanzania.

Satyarupa Shekhar is at the Citizen consumer and civic Action Group (CAG), Chennai, which works to overcome challenges to access to basic services posed by a lack of data and information, while improving transparency and accountability. She works with the city government and other public agencies in Chennai to bring a more data-driven approach to governance and leads the Chennai Data Portal project, which collaborates with government departments to create and use data for decision-making. Prior to joining CAG, she worked at Transparent Chennai, the Centre for Development Finance at IFMR, Democracy Connect, the Indian School of Business, and the World Social Forum 2004. She holds a master's degree in law and economics from the University of Rotterdam, where she was an Erasmus Mundus scholar. She also has a master's in economics from the University of Hyderabad.

Parminder Jeet Singh worked for nearly a decade in the Indian government, where he initiated innovative e-governance projects. In 2001, he coauthored the book, *Government@Net: E-governance Opportunities for India* (SAGE Publications). He was invited to conduct a research study at the INSEAD Business School in France on new ICTs for community and governance institutions. He also worked on many ICTD projects, policy research, and advocacy related to information society issues. At IT for Change, he is the coordinator of a field project funded by the United Nations Development Programme that aims to bring new ICTs to disadvantaged rural women as well as the project Information Society for the South.

Matthew L. Smith has worked at Canada's International Development Research Centre for over a decade, supporting a wide range of research for development projects seeking to harness information and communication technologies for more inclusive societies. During this time, he established himself as a leading thinker on the role of openness in international development, coediting *Open Development: Networked Innovations in International Development* (MIT Press and IDRC, 2013) and *Making Open Development Inclusive: Lessons from IDRC Research* (MIT Press and IDRC, 2020). Aside from his work on openness, he has also published widely on a variety of topics, including e-government, trust, education, the human development and capabilities approach, and the philosophy of social science research. He has a PhD in information systems and an MS in development studies from the London School of Economics and Political Science. He also holds an MS in artificial intelligence from Edinburgh University.

Janaki Srinivasan holds a PhD from the University of California, Berkeley, School of Information and is an assistant professor at the International Institute of Information Technology-Bangalore (IIIT-B). She studies the political economy of information and ICT-focused development initiatives. She is currently working on the role of intermediaries in ICT-based transactions among agricultural actors in India and the role of information determinism in ICT-based initiatives.

Bernd Carsten Stahl specializes in critical research of technology and is director of the Centre for Computing and Social Responsibility at De Montfort University. His research covers philosophical issues arising from the intersections of business, technology, and information. This includes the ethics of ICT and critical approaches to information systems. From 2009 to 2011, he served as coordinator of the EU FP7 research project Ethical Issues of Emerging ICT Applications (ETICA), and from 2012 to 2015 he served as coordinator of the EU FP7 research project Civil Society Organisations in Designing Research Governance (CONSIDER).

John Traxler is professor of digital learning at the Institute of Education at the University of Wolverhampton, United Kingdom. He is one of the pioneers of mobile learning and has been associated with mobile learning projects since 2001. He is coeditor of the definitive books *Mobile Learning: A Handbook for Educators and Trainers* (2005, Routledge) and *Mobile Learning: The Next Generation* (2015, Routledge). He has been responsible for large-scale mobile learning implementations, small-scale mobile learning research interventions, major evaluations, and landscape reviews. He currently works on a large EU project exploring digital learning for continuing professional development in rural areas and is an expert reviewer for EU projects. He has been actively developing innovative approaches to the ethics of mobile and popular digital technologies.

Jean-Paul Van Belle is a professor of information technology and director at the Centre for Information Technology and National Development in Africa at the University of Cape Town. He was head of the Department of Information Technology at the University of Cape Town from 2008 to 2011. He has written over 120 published peer-reviewed articles in the fields of service-oriented architecture, unified communications within businesses, open source software, mobile computing, and information and communications technologies for development.

Marion Walton is currently at the University of Cape Town. Her research explores the connections between media studies and the study of software, digital media, social networks, and games. She has a particular interest in mobile media, developing new research methodologies for the study of interactive media, and user experiences of social and participatory media. Her PhD studies included a period of study at the Centre for the Study of Children, Youth and Media at the Institute of Education, University of London. Her research in human-computer interaction suggests approaches to studying software as a new form of media and confronts the issues of power and regulation of meaning that arise for users of software, particularly those in marginalized contexts.

Yingqin Zheng is senior lecturer at the School of Management, Royal Holloway University of London. She obtained her doctorate from the University of Cambridge as a Gates Scholar. Her research interests include the implications of information and communication technologies in the transformation of organizations and societies. Her work in ICTD explores the contributions of conceptual approaches such as Sen's capability approach. Empirically, she has investigated topics related to health information systems, distributed innovation, social inclusion, social media, and collective action. She currently serves as senior editor for *Information Technology and People* and associate editor for the *Information Systems Journal*.

Index

Note: page numbers followed by *t* and *f* refer to tables and figures respectively. Those followed by n refer to notes, with note number.

Printed in the United States
by Baker & Taylor Publisher Services